SMART AUTOCLAVE CURE OF COMPOSITES

HOW TO ORDER THIS BOOK

BY PHONE: 800-233-9936 or 717-291-5609, 8AM–5PM Eastern Time

BY FAX: 717-295-4538

BY MAIL: Order Department
Technomic Publishing Company, Inc.
851 New Holland Avenue, Box 3535
Lancaster, PA 17604, U.S.A.

BY CREDIT CARD: American Express, VISA, MasterCard

SMART AUTOCLAVE CURE OF COMPOSITES

PETER R. CIRISCIOLI
GEORGE S. SPRINGER
Stanford University

LANCASTER · BASEL

Smart Autoclave Care of Composites
a **TECHNOMIC**® publication

Published in the Western Hemisphere by
Technomic Publishing Company, Inc.
851 New Holland Avenue
Box 3535
Lancaster, Pennsylvania 17604 U.S.A.

Distributed in the Rest of the World by
Technomic Publishing AG

Printed in the United States of America
10 9 8 7 6 5 4 3 2 1

Main entry under title:
Smart Autoclave Care of Composites

A Technomic Publishing Company book
Bibliography: p.

Library of Congress Card No. 90-71382
ISBN No. 87762-802-5

Contents

Preface

Manufacturing composites in an autoclave has largely been an empirical art. Only recently have more rational approaches been applied to the selection of autoclave temperature and pressure histories. As a first step away from empiricism, the manufacturing process was described by equations founded on fundamental engineering principles. These equations were then used to select the autoclave temperature and pressure. This approach was followed by expert systems which, on the basis of preestablished rules, select and control in real time the autoclave conditions. The advantages of rule based expert systems are numerous. They require little information about the material, can be used for complex shaped parts, provide instantaneous feedback control of the autoclave temperature and pressure, and result in high quality parts cured in the shortest time.

In this book we present process models for both thermosetting and thermoplastic matrix composites and an expert system applicable to autoclave curing of thermosetting matrix composites. We recognize that the models and the expert system described in this book are only first steps towards a complete solution of the problem. Nevertheless, the unified presentations given here are deemed appropriate. First, in their present forms the models and the expert system are already practical and useful. Second, the subjects are presented in sufficient detail to acquaint the reader with the rationale behind the models and the rules and with the numerical procedures, the software, and the hardware. Thus, this book should be useful as a reference to practicing manufacturing engineers as well as to workers new to the field.

The authors thank the many individuals and corporations who contributed to this work. The thermosetting model is based on the Loos-Springer formulation. W.I. Lee helped with the thermoplastic model and the expert

system rules.
Q. Wang assisted with curing and computer runs and K. Perry with mechanical testing. A.C. Goetz made available mechanical test facilities at Lockheed Missiles and Space Corporation. R.J. Downs had useful suggestions for the design of the apparatus. Micromet Incorporated and TMI Incorporated provided some of the equipment.

Peter R. Ciriscioli
George S. Springer

Stanford, California
July, 1990

CHAPTER 1

Introduction

When fabricating structures made of fiber reinforced organic matrix composites careful attention must be paid to the manufacturing procedures. The quality of the part may suffer and the cost may become excessive if improper manufacturing processes are used. Therefore, the manufacturing process must be selected carefully to ensure that both the quality and the cost are acceptable. The major process variables which must be selected and controlled are the heat and pressure applied during autoclave curing of thermosetting matrix composites, and the cooling rate and pressure applied during the processing of thermoplastic matrix composites.

Although the aforementioned process variables can be chosen empirically, the empirical approach is undesirable and often impractical. Empirical, trial and error methods are expensive and time consuming, and do not ensure that the resulting processing conditions are optimum. It is far more advantageous and convenient to establish the required process conditions and process variables either by the use of analytical models, or by expert systems.

In this book two models are described, one applicable to the cure of thermosetting matrix composites, the other to the processing of thermoplastics. In addition, a rule based expert system is presented which can be used for real time control of the autoclave temperature and pressure during the cure of thermosetting matrix composites.

PART 1

Curing of Thermosetting Matrix Composites

CHAPTER 2

General Considerations for Curing Thermosetting Matrix Composites

Parts and structures constructed from continuous fiber reinforced thermosetting resin composites are manufactured by arranging the uncured fiber-resin mixture into the desired shape and then curing the material. The curing process is accomplished by exposing the material to elevated temperatures and pressures for a predetermined length of time. The elevated temperatures applied during the cure provide the heat required for initiating and maintaining the chemical reactions in the resin which cause the desired changes in the molecular structure. The applied pressure provides the force needed to squeeze excess resin out of the material, to consolidate individual plies, and to compress voids.

The elevated temperatures and pressures to which the material is subjected are referred to as the cure temperature and the cure pressure. The magnitudes and durations of the temperatures and pressures applied during the curing process (denoted as the cure cycle) affects the following parameters of interest:

a) the temperature inside the composite,
b) the degree of cure of the resin,
c) the resin viscosity,
d) the resin flow, the amount of resin in the composite, and the amount of resin in the bleeder,
e) the changes in the void sizes,
f) the residual stresses and strains in the composite, and
g) the cure time.

By affecting these parameters, the cure cycle has a significant effect on the quality of the finished part. Therefore, the cure cycle must be selected specifically for each application and controlled during the cure process.

Some major considerations in selecting the proper cure cycle for a given composite material are listed in the following table.

Table 2.1

Requirements to be Met During Autoclave Curing
of
Thermosetting Matrix Composites

- The temperature inside the autoclave and at any point inside the cure assembly (composite as well as surrounding layers) must not exceed a preset value at any time during the cure.
- The temperature distribution must be reasonably uniform inside the composite, and there cannot be significant temperature increases due to exotherms induced by chemical reactions.
- The composite must be fully compacted.
- The formation and growth of voids must be minimized.
- The residual ("curing") stresses must be reduced or eliminated.
- Complete and uniform cure must be reached in the shortest time.

The temperature and pressure conditions inside the autoclave are set by adjusting the heater, cooler, and pressure controllers. There are several ways in which the processing conditions and the corresponding autoclave controls (heater, cooler, pressure) can be selected for a given application; 1) by adopting the "manufacturer recommended cure cycle", 2) by using empirical methods, 3) by employing analytic models and, 4) by using expert systems.

Manufacturer recommended cure cycle

The material supplier generally recommends a "cure (pressure-temperature) cycle". Such cycles are based on tests performed on small samples, and do not take into account the geometry of the part. Therefore, the cycle recommended by the manufacturer is inappropriate for most applications, and should not be used without extensive trial beforehand.

Empirical methods

In principle, the processing conditions can be chosen empirically. The schematic of the empirical approach is shown in Figure 2.1. In practice,

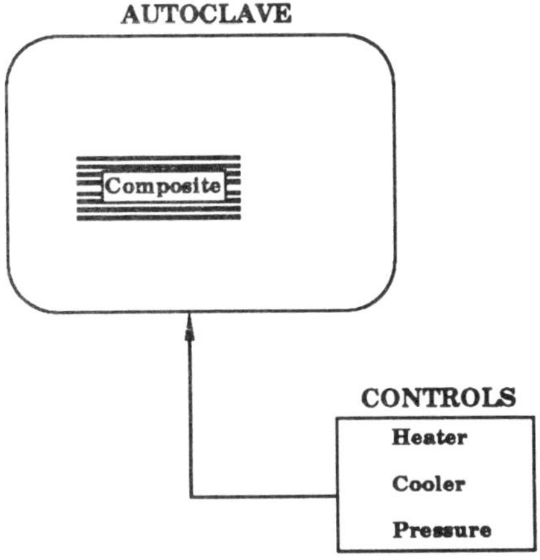

Figure 2.1 The empirical approach for selecting the autoclave temperature and pressure.

empirical approaches are fraught with difficulties. Empirical, trial and error methods are impractical even for thin, small parts, and are impossible for thick, large parts. The obvious difficulties with empirical methods are that a) they are expensive and time consuming, b) their results can not be generalized to parts made of different shapes and different materials, and c) they cannot be used to control the manufacturing process.

Analytic models

The "optimum" processing conditions can be established by analytic models as illustrated by the schematic in Figure 2.2. Such models have already been developed for autoclave curing [1-7] and filament winding [8-11] of thermoset matrix composites and for processing of thermoplastic composites [12-16]. Although analytic models may provide the processing conditions for certain problems, they are beset by difficulties. Most notably, analytic models a) require extensive material data which often are difficult to measure, b) can only be applied to simple geometries, c) do not take into account "batch to batch" variations in material properties, and, last but not least, d) cannot be used to control the manufacturing process.

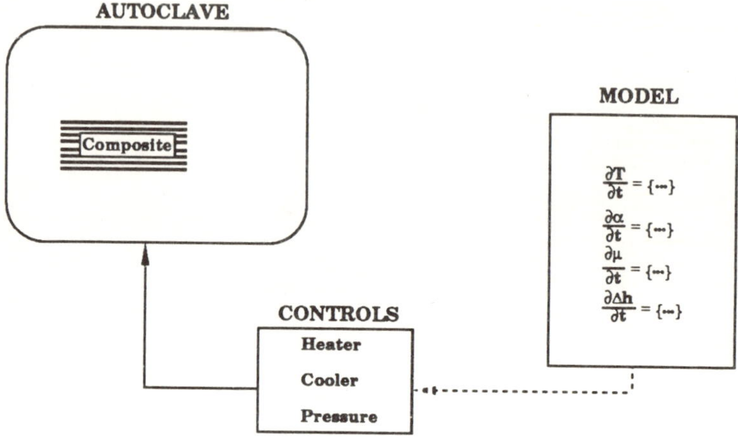

Figure 2.2 The analytical model approach for selecting the autoclave temperature and pressure.

Expert systems

The shortcomings of the aforementioned three methods can best be overcome by the use of expert systems, which provide suitable means for both selecting and controlling in real time the processing conditions [17 - 19]. With such systems certain variables (e.g. temperature, pressure, ionic conductivity, dimension) are monitored inside the material. The trends (but not the absolute values) of the data are examined. Preestablished rules are then applied to the data and, on this basis, the heater, the cooler, and the pressure controls are adjusted (Figure 2.3).

Expert systems do not require prior knowledge of the material properties, and are not limited to simple shapes. Thus, they offer excellent means for selecting and controlling the processing conditions, such that all the required conditions are met (Table 2.1); namely the temperatures remain within prescribed limits, full compaction and complete cure are achieved, and the void content and residual strains are minimized.

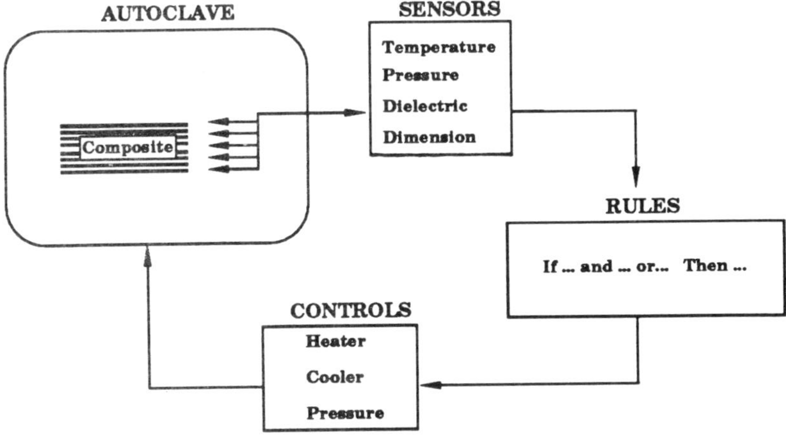

Figure 2.3 The expert system scheme for selecting the autoclave temperature and pressure.

In this book two means are described for selecting the autoclave cure cycle for thermosetting matrix composites. The first is an analytic model, the second, an expert system.

CHAPTER 3

A Model of the Autoclave Cure Process

In this chapter a model, first presented by Loos and Springer [1, 2], is described. This model yields the following parameters during cure:

a) the temperature inside the composite as functions of position and time,

b) the pressure inside the composite as functions of position and time,

c) the degree of cure of the resin as functions of position and time,

e) the number of compacted prepreg plies as functions of position and time,

f) the amount of resin in the bleeder as a function of time,

g) the thickness and the mass of the composite as a function of time,

h) the void size as a function of location and time, and,

i) the residual stresses and strains in each ply after cure.

The model providing the above-mentioned information is developed below in four parts. The first part of the model, referred to as the "thermo-chemical model", yields the temperature, the degree of cure, and the viscosity. The second part ("flow model") gives the pressure, the resin flow out of the composite, and the resin contents of the composite and the bleeder. The third part ("void model") gives the void sizes. The fourth part ("stress model") yields the residual stresses and strains. Details of the models are presented subsequently. First, a brief description of the problem is given.

3.1 PROBLEM STATEMENT

We consider a fiber reinforced epoxy matrix composite of initial thickness L_i constructed from unidirectional continuous fiber "prepreg" tape (Figure 3.1). An absorbent material (referred to as a "bleeder") is placed on one (or on both sides) side of the composite. The thickness of the bleeder is L_b. The composite-bleeder system is placed on a metal tool plate ready for processing. A sheet of nonporous teflon release cloth is placed between the composite and the bleeder to prevent sticking. A metal plate is placed on top of the bleeder, and an air breather is added when curing is done in an autoclave. Restraints (called "dams") are also mounted around the prepreg to prevent lateral motion and to minimize resin flow parallel to the tool plate and through the edges. Finally, a plastic sheet ("vacuum bag") is placed around the entire assembly when vacuum is applied during the cure. Here, we are concerned only with the composite-bleeder system illustrated in Figure 3.2, because the additional components (vacuum bag, air breather, teflon sheets, etc.) have no direct effect on the model.

Initially (time < 0), the resin is uncured and the bleeder contains no resin. Starting at time $t = 0$, the composite-bleeder system is exposed to a known temperature T_o. The cure temperature T_o may be the same or may be different on the two sides of the composite-bleeder system. At some time t_p, ($t_p \geq 0$) a known pressure P_o is applied to the system. Both the cure temperature T_o and the cure pressure P_o may vary with time in an arbitrary manner. The

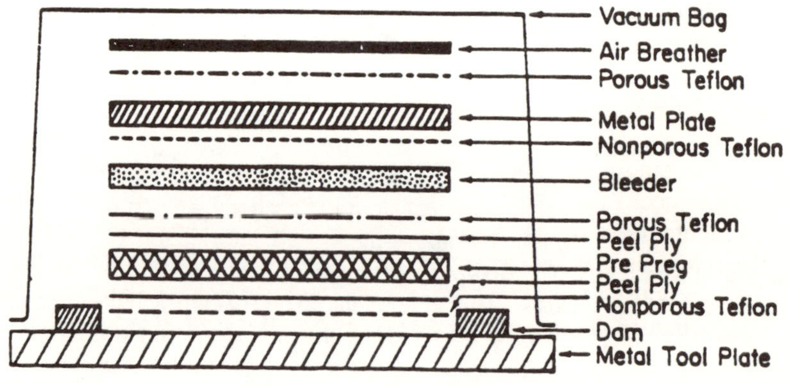

Figure 3.1 Schematic of the prepreg lay-up.

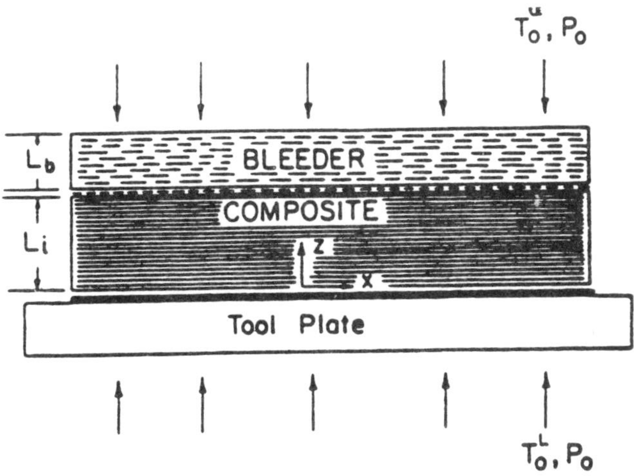

Figure 3.2 Geometry of the composite-bleeder system.

objective is to determine the parameters listed in points *a-i* previously.

In formulating the model, resin is allowed to flow in the directions both perpendicular and parallel to the plane of the composite. Resin flow in the plane of the composite is allowed only in the direction parallel to the fibers. In order to emphasize the concepts and the solution methods, the properties in the plane of the composite are taken to be constant. However, the model and the method are general and could be extended to complex geometries. It is also noted that while the analysis is presented for composites made of continuous fiber, unidirectional tape, the model can also be applied to composites made of woven fabric.

3.2 THERMO-CHEMICAL MODEL

The temperature distribution, the degree of cure of the resin, and the resin viscosity inside the composite depend on the rate at which heat is transmitted from the environment into the material. The temperature inside the composite can be calculated using the law of conservation of energy together with an appropriate expression for the cure kinetics. By neglecting energy transfer by convection, the energy equation may be expressed as

$$\rho C \frac{\partial T}{\partial t} = \frac{\partial}{\partial z}\left(k \frac{\partial T}{\partial z}\right) + \rho \dot{H} \tag{3.1}$$

where ρ and C are the density and specific heat of the composite, k is the thermal conductivity in the direction perpendicular to the plane of the composite, and T is the temperature. z is shown in Figures 3.2 and 3.3. $\dot{H}$ is the rate of heat generated by chemical reactions

$$\dot{H} = RH_U \tag{3.2}$$

H_U is the total or ultimate heat of reaction during cure and R is the reaction or cure rate. The degree of cure of the resin (denoted as the degree of cure, α) is defined as

$$\alpha \equiv \frac{H(t)}{H_U} \tag{3.3}$$

$H(t)$ is the heat evolved from the beginning of the reaction to some intermediate time, t. For an uncured material $\alpha = 0$, and for a completely cured material α approaches unity. By differentiating Eq. 3.3 with respect to time, the following expression is obtained.

$$\dot{H} = \frac{d\alpha}{dt} H_U \tag{3.4}$$

A comparison of Eqs. 3.2 and 3.4 shows that, in this formulation, $d\alpha/dt$ is the reaction or cure rate. If diffusion of chemical species is neglected, the degree of cure at each point inside the material can be calculated in the following way

$$\alpha = \int_0^t \left(\frac{d\alpha}{dt} \right) dt \tag{3.5}$$

In order to complete the model, the dependence of the cure rate on the temperature and on the degree of cure must be known. This dependence may be expressed symbolically as

$$\frac{d\alpha}{dt} = f(T, \alpha) \tag{3.6}$$

The functional relationship in Eq. 3.6, along with the value of the heat of reaction H_u, can be determined experimentally by the procedures described in references [20, 21] and in Appendix A.

The density ρ, specific heat C, heat of reaction H_u, and thermal conductivity k depend on the instantaneous, local resin and fiber contents of each ply. The properties ρ, C, and H_u can be calculated by the rule of mixtures [22], the thermal conductivity of the prepreg by the approximate formula developed by Springer and Tsai [23] (Appendix B).

Solutions to Eqs. 3.1 and 3.4-3.6 can be obtained once the initial and boundary conditions are specified. The initial conditions require that the temperature and degree of cure inside the composite be given before the start of the cure (time < 0). The boundary conditions require that the temperatures on the top and bottom surface of the composite be known as a function of time during cure (time ≥ 0). Accordingly, the initial and boundary conditions corresponding to Eqs. 3.1 and 3.4-3.6 are

Initial conditions:

$$\left.\begin{aligned} T &= T_i(z) \\ \alpha &= \alpha_i \end{aligned}\right\} \begin{aligned} 0 \le z \le L \\ t < 0 \end{aligned} \qquad (3.7)$$

T_i and α_i are the initial temperature and degree of cure in the composite.

Boundary conditions:

$$\left.\begin{aligned} T &= T_L(t)_{z=0} \\ T &= T_u(t)_{z=L} \end{aligned}\right\} t \ge 0 \qquad (3.8)$$

where T_u and T_L are the temperatures on the top ($z = L$) and bottom ($z = 0$) surfaces of the composite, respectively (Figure 3.2). Solutions to Eqs. 3.1 and 3.4-3.8 yield the temperature T, the cure rate $d\alpha/dt$, and the degree of cure α as functions of position and time inside the composite.

Once these parameters are known, the resin viscosity can be calculated, provided a suitable expression relating resin viscosity to the temperature and degree of cure is available. If the resin viscosity is assumed to be independent of shear rate, then the relationship between viscosity, temperature, and degree of cure can be represented in the form [20, 21]

$$\mu = g(T, \alpha) \tag{3.9}$$

The manner in which the relationship between viscosity, temperature, and degree of cure can be established is described in Appendix A.

In the foregoing, attention is focused on the composite. In practice, the heat transfer through the entire cure assembly (see Figure 3.1) must be calculated to establish the composite surface temperatures T_u and T_L. These calculations are performed by using the energy equation (Eq. 3.1) without the heat generation term $\rho\dot{H}$, and by specifying either the temperatures (T_o^u, T_o^L) or the heat fluxes on the top and bottom surfaces of the cure assembly (Figure 3.3). The calculations then proceed in the customary manner for heat conduction through multilayered media.

3.3 RESIN FLOW MODEL

At some time t_p ($t_p \geq 0$), pressure is applied to the composite-bleeder system (Figure 3.3). As a result of the pressure, resin flows from the composite into the bleeders. Resin flow in the direction parallel to the plane of the composite can be neglected if a) both the width and the length of the composite are large compared to the thickness L, and b) if restraints are placed around the sides of the composite. This situation is generally encountered in practice. When modelling the curing process of systems where the aforementioned conditions are met, only resin flow normal to the tool plate needs to be taken

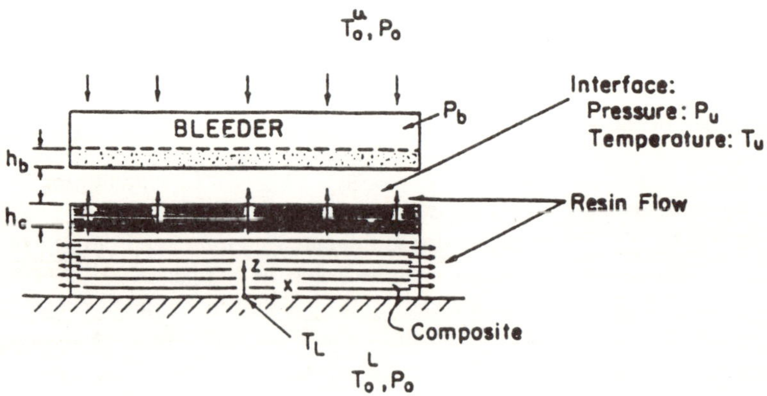

Figure 3.3 Resin flow model.

into account. However, under some conditions, resin flow along the fibers cannot be prevented. This situation may occur when the length of the composite is similar to the thickness. Under these circumstances, resin flow both normal and parallel to the tool plate takes place simultaneously. The model must then consider resin flow in both directions. In the model that is developed below, resin flow both normal and parallel to the tool plate is taken into account.

Before the resin flow model is established, the behavior of the prepreg plies during the squeezing action (cure pressure application) is examined.

The resin flow process normal to the tool plate was demonstrated by Springer [24] to occur by the following mechanism. As pressure is applied, the first (top) ply ($n_s = 1$, Figure 3.4) moves toward the second ply ($n_s = 2$), while resin is squeezed out from the space between the plies. The resin seeps through the fiber bundles of the first ply. When the fibers in the first ply get close to the fibers in the second ply, the two plies move together toward the third ply, squeezing the resin out of the space between the second and the third plies. This sequence of events is repeated for the subsequent plies. Thus, the interaction of the fibers proceeds down the prepreg in a wavelike manner.

Note that there are essentially two regions in the composite. In region 1, the plies are squeezed together and contain no excess resin, while in region 2 the plies have not moved and have the original resin content. Some compacting of the fibers within the individual plies may also occur but, as a first approximation, this effect is neglected here.

It is noted that according to this model there is a pressure drop only across those plies through which resin flow takes place. The pressure is

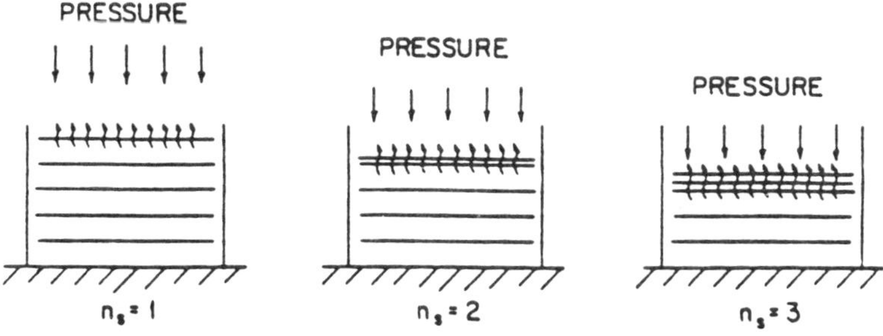

Figure 3.4 Illustration of the resin flow process normal to the tool plate.

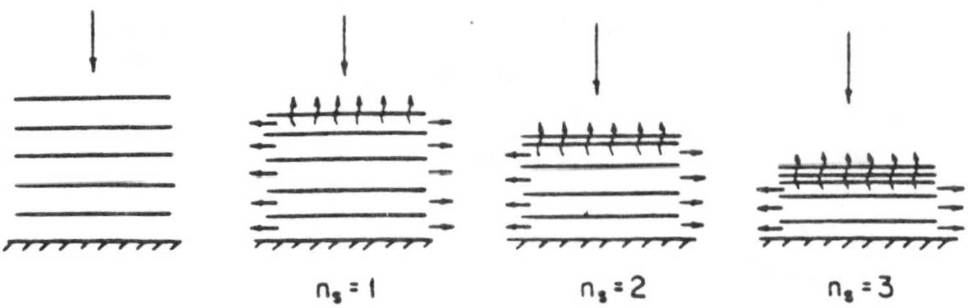

Figure 3.5 Simultaneous resin flow both normal and parallel to the tool plate.

constant (and equal to the applied pressure P_o) across the remaining layers of the prepreg.

When there is resin flow in both the normal and parallel directions, resin is squeezed out from between every ply continuously, as long as there is excess resin between adjacent plies. In this case, the thickness between different plies varies and changes with time, as illustrated in Figure 3.5.

Although the resin flow in the normal and parallel directions are related, to facilitate the calculations in the model, the two phenomena are decoupled. Hence, separate models are described below for the resin flow in directions normal and parallel to the tool plate.

The model developed predicts changes in the dimensions of the composite only due to changes in the resin content. Shrinkage due to changes in the molecular structure of the resin during cure is not considered.

Resin Flow Normal to the Tool Plate

Owing to the complex geometry, the equations describing the resin flow through the composite normal to the tool plate (z direction) and into the bleeder cannot be established exactly. Nevertheless, an approximate formulation of the problem is feasible by treating the resin flow through both the composite and bleeder as flow through porous media. Such an approach was proposed by Bartlett [25] and by Loos and Springer [1, 2] for resin flow in continuous fiber prepregs. In the model, inertia forces are considered to be negligible compared to viscous forces. Then, at any instant of time, the resin velocities in the prepreg and in the bleeder may be represented by Darcy's law

$$V = -\frac{S}{\mu}\frac{dP}{dz} \tag{3.10}$$

where S is the apparent permeability, μ is the viscosity, and dP/dz is the pressure gradient. The law of conservation of mass (together with Eq. 3.10) gives the following expression for the rate of change of mass M in the composite

$$\frac{dM}{dt} = -\rho_r A_z V_z = -\rho_r A_z S_c \frac{P_c - P_u}{\int_0^{h_c} \mu dz} \tag{3.11}$$

where ρ_r is the resin density, V_z and A_z are the velocity and the cross sectional area perpendicular to the z axis, h_c is the thickness of the compacted plies, i.e. the thickness of the layer through which resin flow takes place (Figure 3.3). P_u is the pressure at the interface between the composite and the bleeder. The subscript c refers to conditions in the composite at position h_c. Accordingly, P_c is the pressure at h_c and is the same as the applied pressure ($P_c = P_0$).

At any instant of time, the resin flow rate into the bleeder is

$$\rho_r A_z V_z = \rho_r A_z V_b \tag{3.12}$$

The temperature, and hence the viscosity, of the resin inside the bleeder is assumed to be independent of position (but not of time). Thus, Eqs. 3.10 and 3.12 yield

$$\rho_r A_z V_z = \rho_r A_z \frac{S_b}{\mu_b}\frac{P_u - P_b}{h_b} \tag{3.13}$$

where h_b is the instantaneous depth of resin in the bleeder. The subscript b refers to conditions in the bleeder. In developing the above expressions, the pressure drop across the porous teflon sheet between the bleeder and the composite is neglected.

Noting that the mass of the fibers in the composite remains constant, Eqs. 3.11-3.13 may be rearranged to yield the following expression for the rate of change of resin mass in the composite

$$\frac{dM_r}{dt} = \frac{-\rho_r A_z S_c}{\int_0^{h_c} \mu dz} \left(\frac{P_o - P_b}{1 + G(t)} \right) \tag{3.14}$$

The parameter $G(t)$ is defined as

$$G(t) \equiv \frac{S_c}{S_b} \frac{\mu_b h_b}{\int_0^{h_c} \mu dz} \tag{3.15}$$

M_r is the mass of resin in the composite at any instant of time. The mass of resin that leaves the composite and enters the bleeder in time t is

$$M_T = \int_0^t \frac{dM_r}{dt} dt \tag{3.16}$$

The instantaneous resin depth in the bleeder is related to the mass of resin that enters the bleeder by the expression

$$h_b = \frac{1}{\rho_r \phi_b A_z} \int_0^t \frac{dM_r}{dt} dt \tag{3.17}$$

where ϕ_b is the porosity of the bleeder and represents the volume (per unit volume) which can be filled by resin. The thickness of the compacted plies is

$$h_c = n_s h_1 \tag{3.18}$$

where h_1 is the thickness of one compacted prepreg ply and n_s is the number of compacted prepreg plies. The value of n_s varies with time, depending on the amount of resin that has been squeezed out of the composite.

Equations 3.11-3.18 are the relationships needed for calculating the resin flow normal to the tool plate.

Resin Flow Parallel to the Tool Plate

In principle, in the plane of the composite, resin may flow along the fibers and in the direction perpendicular to the fibers. In practice, resin flow perpendicular to the fibers is small because of a) the resistance created by the fibers and b) the restraints placed around the edges of the composite. If such restraints were not provided, fiber spreading ("wash out") would occur, resulting in a non-uniform distribution of fibers in the composite. Therefore, in this section, only resin flow along the fibers is considered.

It is assumed that resin flow along the fibers and parallel to the tool plate can be characterized as viscous flow between two parallel plates separated by a distance d_n ("channel flow", Figure 3.6). The distance d_n separating the plates is small compared to the thickness of the composite ($d_n < L$). The variation in resin properties across and along the channel are taken to be constant. The pressure drop between the center of any given channel and the

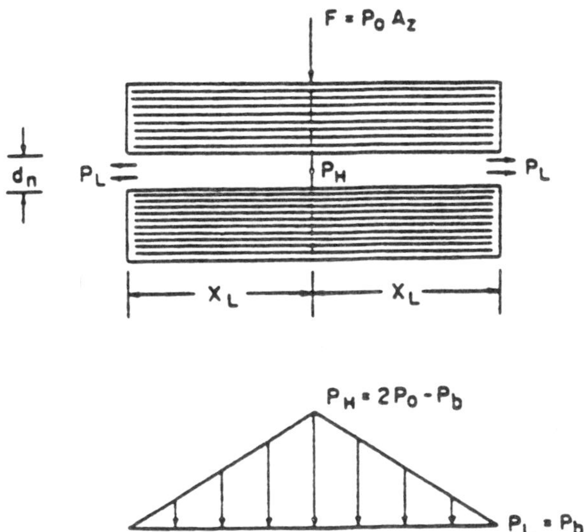

Figure 3.6 Geometry of the resin flow model parallel to the tool plate.

channel exit (P_H - P_L, Figure 3.6) can then be expressed as [26]

$$\frac{2(P_H - P_L)}{\rho_r \left(V_x^2\right)_n} = \lambda \frac{X_L}{d_n} \tag{3.19}$$

where $(V_x)_n$ is the average resin velocity in the channel. X_L is the channel length. The subscript n refers to the channel located between the n and the n -1 prepreg plies (i.e., beneath the fiber bundles of prepreg ply n). The thickness of the nth channel is calculated by assuming that a) there is one channel per ply and b) all the excess resin is contained in the channel. Accordingly, the thickness of the channel is given by the following expression

$$d_n = \frac{M_n}{\rho_n A_z} - \frac{M_{com}}{\rho_{com} A_z} \tag{3.20}$$

The mass M_n and density ρ_n of prepreg ply n can be calculated by the rule of mixtures (Appendix B). M_{com} and ρ_{com} are the mass and the density of a compacted prepreg ply, respectively. The technique used to determine M_{com} is described in the next section, along with an appropriate expression for determining ρ_{com}. For laminar flow between parallel plates, λ is defined as

$$\lambda \equiv \frac{(1/B)\mu_n}{\rho_r (V_x)_n d_n} \tag{3.21}$$

where μ_n is the viscosity of the resin in the channel. Substitution of Eq. 3.21 into Eq. 3.19 yields the following expression for the average velocity in the channel

$$(V_x)_n = 2B \frac{d_n^2}{\mu_n} \frac{(P_H - P_L)}{X_L} \tag{3.22}$$

where B is a constant which must be determined experimentally. The resin mass flow rate is

$$(\dot{m}_{rx})_n = \rho_r A_x (V_x)_n \tag{3.23}$$

where A_x is the cross sectional area defined as the product of the channel width W and thickness d_n.

The law of conservation of mass, together with Eqs. 3.22-3.23, gives the following expression for the rate of change of mass in the nth prepreg ply

$$\frac{d(M_r)_n}{dt} = -2(\dot{m}_{rx})_n = -2B\frac{d_n^3}{\mu}\rho_r W\frac{(P_H - P_L)}{X_L} \quad (3.24)$$

The amount of resin leaving the nth prepreg ply in time t is

$$(M_E)_n = \int_0^t \frac{d(M_r)_n}{dt}dt \quad (3.25)$$

The total amount of resin flow parallel to the tool plate can be determined by summing Eq. 3.25 over all the plies containing excess resin

$$M_E = \sum_{n=1}^{N-n_s}(M_E)_n \quad (3.26)$$

where N is the total number of prepreg plies.

The pressure at the centerline of the channel P_H can be estimated from the force balance applied along the boundaries of the channel. Assuming that the pressure gradient in the x direction is linear, and that the centerline pressure P_H is the same in each ply, the pressure distribution of each channel may be expressed as

$$P = \left(\frac{P_L - P_H}{X_L}\right)x + P_H \quad (3.27)$$

where P_L is the pressure at the exit of the channel and is assumed to be equal to the pressure of the environment surrounding the composite P_b. A force balance applied along the channel surface gives (Figure 3.6)

$$F = \int_A PdA = 2W\int_0^{X_L} Pdx \quad (3.28)$$

F is the applied force which can be related to the cure pressure P_o as

$$F = P_o A_z = 2P_o W X_L \tag{3.29}$$

Equations 3.27-3.29 yield the centerline pressure

$$P_H = 2P_o - P_b \tag{3.30}$$

Equations 3.19-3.30 are the relationships needed for calculating the resin flow in the direction along the fibers.

Total Resin Flow

The total resin flow out of the composite in time t is the sum of the resin flows both normal and parallel to the tool plate. The law of conservation of mass gives the following expression for the total rate of change of mass M in the composite

$$\frac{dM}{dt} = -\left[\dot{m}_{rz} + 2\sum_{n=1}^{N-n_s}(m_{rx})_n\right] \tag{3.31}$$

where m_{rz} and m_{rx} are the resin mass flow rates normal to the tool plate (z direction) and parallel to the tool plate (x direction), respectively.

The total mass of the composite at time t is

$$M = M_i - M_T - M_E \tag{3.32}$$

where M_T and M_E are defined by Eqs. 3.16 and 3.26, and M_i is the initial mass of the composite. The composite thickness at time t is

$$L = \frac{M}{\rho 2 X_L W} \tag{3.33}$$

where ρ is the density of the composite.

Modifications to the above flow model have been proposed by Gutowski and his co-workers [3, 4] and by Kardos et al. [5].

3.4 VOID MODEL

Void nuclei may be formed either by mechanical means (e.g. air or gas bubble entrapment, broken fibers) or by homogeneous or heterogeneous nucleation. Once a void is established, its size may change due to three effects: a) changes in vapor mass inside the void caused by vapor transfer across the void-prepreg interface, b) changes in pressure inside the void due to changes in temperature and pressure in the prepreg, and c) thermal expansion (or shrinking) due to temperature gradients in the resin. Models which take into account the first two of these effects, namely vapor transfer and changes in temperature and pressure have been developed by Loos and Springer [1, 2] and by Kardos et al [6, 27]. Here, details of the Loos-Springer model are presented.

A spherical nucleus of diameter d_i is assumed to be at a given location in the prepreg. The nucleus is filled with water vapor resulting from the humid air surrounding the prepreg during lay-up. The partial pressure of the water vapor in the nucleus PP_{wi} is related to the relative humidity RH by the expression

$$PP_{wi} = (RH)(P_{wga}) \tag{3.34}$$

where P_{wga} is the saturation pressure of the water vapor at the ambient temperature. From the known values of the initial partial pressure PP_{wi} and the initial nucleus volume, the initial mass m_{wi} and the initial concentration of the water vapor in the nucleus can be determined.

During the cure, the volume of the void changes because a) water and other types of molecules are transported across the void-prepreg interface, and b) the pressure changes at the location of the void. For a spherical void of diameter d, the total pressure inside the void P_v is related to the pressure in the prepreg surrounding the void P by the expression

$$P_v - P = \frac{4\sigma}{d} \tag{3.35}$$

σ is the surface tension between the resin and the void. P_v is the total pressure inside the void and is the sum of the partial pressures of the air and the different types of vapors present in the void. In the model outlined below, it is assumed that only water-vapor is transported through the void-prepreg interface. However, other types of vapors can readily be included in the calculations, as described by Springer [1].

The pressure inside the void is

$$P_v = PP_w + PP_{air} \tag{3.36}$$

PP_w and PP_{air} are the partial pressures of the water vapor and the air in the void. The partial pressure is a known function of the temperature, mass and the void diameter

$$\begin{aligned} PP_{air} &= f(T, m_{air}, d) \\ PP_w &= f(T, m_w, d) \end{aligned} \tag{3.37}$$

Thus, if the pressure in the prepreg around the void, the temperature inside the void (taken to be the same as the temperature of the prepreg at the void location), and the mass of vapor in the void are known, the partial pressure, the total pressure, and the void diameter can be calculated from Eqs. 3.35-3.37. The temperature and the pressure are given by the thermochemical-resin-flow models. The air mass in the void is taken to be constant. Thus, it remains to evaluate the mass of water vapor in the void as a function of time. The change in water vapor mass may be calculated by assuming that the vapor molecules are transported through the prepreg by Fickian diffusion. Fick's law gives

$$\frac{dc}{dt} = D\left(\frac{\partial^2 c}{\partial r^2} + \frac{2}{r}\frac{\partial c}{\partial r}\right) \tag{3.38}$$

c is the water vapor concentration at a radial coordinate r with $r = 0$ at the center of the void. D is the diffusivity of the water vapor through the resin in the r direction.

Initially ($t < 0$) the vapor is taken to be distributed uniformly in the prepreg at the known concentration c

$$c = c_i \quad at \ r \geq d_i / 2 \quad t < 0 \tag{3.39}$$

At times $t \geq 0$ the vapor concentration at the void-prepreg interface must be specified. By denoting the concentration at the prepreg surface by the subscript m, we write

$$\left.\begin{array}{ll} c = c_m & at\ r = d/2 \\ c = c_i & at\ r \rightarrow \infty \end{array}\right\} t \geq 0 \tag{3.40}$$

The second of the above expressions reflects the fact that the concentration remains unchanged at a distance far from the void. The surface concentration is related to the maximum saturation level M_m in the prepreg by the expression

$$c_m = \rho M_m \tag{3.41}$$

The value of M_m can be determined experimentally for each vapor-resin system [28].

Solutions of Eqs. 3.38-3.41 yield the vapor concentration as a function of position and time $c = f(r,t)$. The mass of vapor transported in time t through the surface of the void is

$$m_T = -\int_0^t \pi d^2 D \left(\frac{\partial c}{\partial r}\right)_{r=d/2} dt \tag{3.42}$$

The mass of vapor in the void at time t is

$$m = m_{wi} - m_T \tag{3.43}$$

The initial mass of water vapor in the void is known, as was discussed previously.

Solutions to Eqs. 3.34-3.43 give the void size and the pressure inside the void as functions of time, for a void of known location and initial size.

3.5 STRESS MODEL

The procedure for calculating the residual stresses can be found in texts such as Tsai and Hahn [29] and Jones [30]. The development presented here is along the line given by Tsai and Hahn. Only the outlines of the calculation procedure are included here. For details the reader is referred to reference [29].

As a starting point, we consider a [0/90] laminate cooled from the cure temperature to room temperature T_a (Figure 3.7). If the two plies were unconnected the 0 degree ply would deform by e_1 and the 90 degree ply would deform by e_2. Furthermore, the plies would be stress free. In practice, the two plies are connected and the deformation of both plies is the same e_o. The subscript zero indicates that the strain across the laminate thickness is constant. This strain gives rise to the residual stresses in the 0 and 90 degree plies. This specific example may now be generalized to a symmetric laminate with arbitrary ply orientation. For such a laminate the residual stress in any given ply is

$$\sigma_i = Q_{ij}\left(e_{oj} - e_j\right) \tag{3.44}$$

Q_{ij} is the modulus as defined by Tsai and Hahn [29]. The relationship between Q_{ij} and the engineering constants can be found in reference [29]. The strain e_{ij} may be computed from the expression

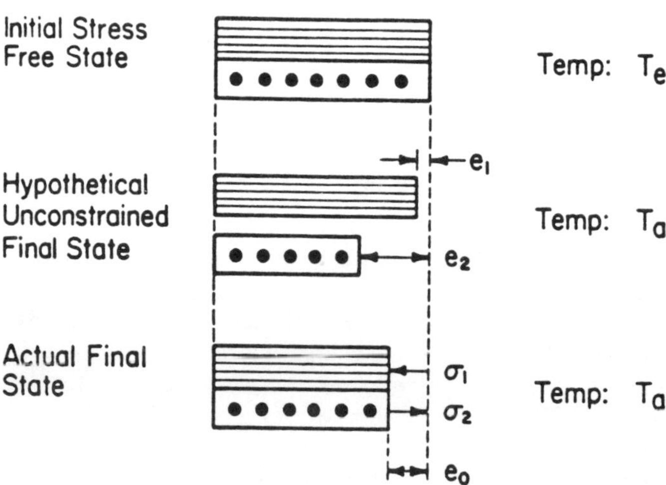

Figure 3.7 Illustration of the buildup of residual streses after curing.

$$e_j = \alpha_j (T_e - T_a) \tag{3.45}$$

where T_e is the temperature in the ply under consideration at the time the resin solidifies and T_a is the ambient temperature. α_j is the thermal expansion coefficient. The laminate curing strain e_{0j} is

$$e_{oj} = a_{ij} \int_0^L Q_{ij} e_j dz \tag{3.46}$$

L is the thickness of the laminate, z is the coordinate perpendicular to the plane of the laminate. a_{ij} is the in-plane compliance of a symmetric laminate as defined by Tsai and Hahn.

Equations 3.43-3.46 comprise the equations needed for calculating the residual stresses in each ply.

3.6 METHOD OF SOLUTION

Solutions to the thermo-chemical, flow, void, and stress models must be obtained by numerical methods. A computer code (designated as "CURE") suitable for generating solutions was developed by Loos and Springer [2].

Solutions of the model (and the corresponding computer code) require that the parameters listed in Tables 3.1-3.3 be specified. The parameters pertaining to the geometry, along with the initial and boundary conditions are specified by the user of the prepreg. The properties of the prepreg, the fiber, the resin, and the bleeder cloth are either specified by the manufacturer or can often be found in the published literature. Parameters for two material systems, namely for Fiberite T300/976 and Hercules AS4/3501-6 are given in Table 3.4. Some input items (the resin content of one compacted ply, the compacted ply thickness, the prepreg apparent permeability normal to the plane of the composite, the prepreg flow coefficient parallel to the fibers and the surface tension at the resin-void interface) in Tables 3.1-3.3 are generally unknown. In the following, a brief description is given of the methods which can be used to determine these parameters.

The compacted prepreg ply thickness and the compacted prepreg ply resin content can be determined by constructing a thin (4 to 16 ply) composite panel. The panel is cured by employing a cure cycle which insures that all the excess resin is squeezed out of every ply in the composite (i.e., all plies

Table 3.1

Parameters Required for Solutions to the Thermo-Chemical and Resin Flow Models
Geometry
Length of the composite
Width of the composite
Number of plies in the composite
Initial and Boundary Conditions
Initial temperature distribution in the composite
Initial degree of cure of the resin in the composite
Cure temperature as a function of time
Cure pressure as a function of time
Pressure in the bleeder
Prepreg Properties
Initial thickness of one ply
Initial resin mass fraction
Resin content of one compacted ply
Compacted ply thickness
Apparent permeability normal to the plane of the composite
Flow coefficient parallel to the fibers
Resin Properties
Density
Specific heat
Thermal conductivity
Heat of reaction
Relationship between the cure rate, the temperature, and the degree of cure
Relationship between the viscosity, the temperature, and the degree of cure
Fiber Properties
Density
Specific heat
Thermal conductivity

Table 3.1, *continued*

Bleeder properties
- Apparent permeability
- Porosity

Table 3.2

Parameters Required for Solutions to the Void Model
(These paramaters are in addition to those in Table 3.1)

Initial and Boundary Conditions
- Initial void size
- Initial void location
- Initial water concentration in the prepreg
- Ambient relative humidity
- Ambient temperature

Resin Properties
- Expression relating the relative humidity to the maximum saturation level of water vapor in the resin
- Diffusivity of water vapor through the resin
- Surface tension at the resin-void interface

Table 3.3

Parameters Required for Solutions to the Stress Model
(These paramaters are in addition to those in Table 3.1)

Properties of the composite
- Longitudinal and transverse Young's modulii
- Longitudinal and transverse Poisson's ratios
- Longitudinal and transverse shear modulii
- Longitudinal and transverse thermal expansion coefficients

Stacking Sequence
- Orientation of each ply

Environment
- Ambient temperature

Table 3.4

Degree of Cure and Viscosity
of Fiberite 976 and Hercules 3501-6 Resins

	976 $0<\alpha<1.0$		3501-6 $0<\alpha\leq 0.3$	3501-6 $0.3<\alpha\leq 1.0$
A_1 (min^{-1})	2.64 x 10^5		2.101 x 10^9	1.960 x 10^5
A_2 (min^{-1})	4.23 x 10^5		-2.014 x 10^9	0
ΔE_1 (J/mol)	6.25 x 10^4		8.07 x 10^4	5.66 x 10^4
ΔE_2 (J/mol)	5.68 x 10^4		7.78 x 10^4	
B	1.0		0.47	0
a	1.03		1.0	0
b	1.22		1.0	0
c	0		1.0	1.0
d	0		1.0	1.0
H_u	530		474	474
H_T/H_u	$0.0044T-1.1$	$T<480°K$	1.0	1.0
	1.0	$T\geq 480°K$		
μ (Pa S)	1.06 x 10^{-6}		7.93 x 10^{-14}	7.93 x 10^{-14}
U (J/mol)	3.76 x 10^4		9.08 x 10^4	9.08 x 10^4
κ	18.8		14.1	14.1

$$\frac{d\alpha}{dt}=\frac{H_T}{H_u}\left(K_1+K_2\beta^a\right)\left(B-\beta\right)^b\left(1-\beta^d\right)^c$$

$$\beta=\frac{H_T}{H_u}\int_0^t\frac{d\alpha}{dt}dt$$

$$K_{1,2}=A_{1,2}\exp(-\Delta E/RT)$$

$$\mu=\mu_\infty\exp(U/RT+\kappa\alpha)$$

are consolidated $n_s = N$). The total mass of the composite M is measured after cure. The resin content of one compacted prepreg ply $(M_r)_{com}$ is related to the composite mass by the expression

$$(M_r)_{com} = \frac{M}{N} - M_f \tag{3.47}$$

M_f is the fiber mass of one prepreg ply, and N is the total number of plies in the composite. The compacted prepreg ply thickness h_1 is

$$h_1 = \frac{M / N}{\rho_{com} A_z} \tag{3.48}$$

where ρ_{com} is compacted ply density which can be estimated by the rule of mixtures given in Appendix B.

The apparent permeability of the prepreg normal to the fibers S_c can be determined by curing a thin (4 to 8 ply) composite specimen for a predetermined length of time. During the cure, the resin squeezed out through the plane of the composite normal to the tool plate is collected in the bleeder placed on the top of the composite. After the cure is terminated, the amount of resin in the bleeder (i.e., the resin flow into the bleeder) is determined by measuring the difference between the original bleeder weight (mass) and the final weight (mass) of the resin-soaked bleeder. An initial value for the apparent permeability is estimated, and the resin flow normal to the fibers is calculated using the flow model. The value of the permeability is adjusted and the calculations are repeated until the calculated and measured resin flows match.

The flow coefficient of the prepreg parallel to the fibers (B) can be estimated from the following procedure. A thick composite (approximately 30-60 plies thick) is cured for a predetermined length of time. Resin squeezed out from between the individual plies (parallel to the fibers) is collected by bleeders placed around the edges of the composite. The resin flow through the edges is determined by measuring the difference between the original weight of the "edge" bleeders and the final weight of the resin soaked bleeders. Assuming a value for the flow coefficient, the resin flow parallel to the fibers is calculated using the flow model. The value of the flow coefficient is adjusted and the calculations are repeated until the measured resin flow matches the calculated resin flow.

The surface tension at the void-resin interface may be approximated by the surface tension of water.

3.7 SELECTION OF THE AUTOCLAVE TEMPERATURE AND PRESSURE

The CURE computer code can readily be used to determine the appropriate cure cycle for a given application. The procedure is straightforward.

1) An arbitrarily chosen cure cycle (cure temperature and cure pressure) is entered into the code.
2) The results shown in Figure 3.8 and Table 3.5 are calculated and plotted.
3) The results are examined to see if all the conditions specified in Table 2.1 are satisfied.
4) If the conditions in Table 2.1 are not satisfied, a new cure cycle is selected, and the procedure is repeated until all requirements are met. Generally, the required cure cycle is obtained within a few iterations.

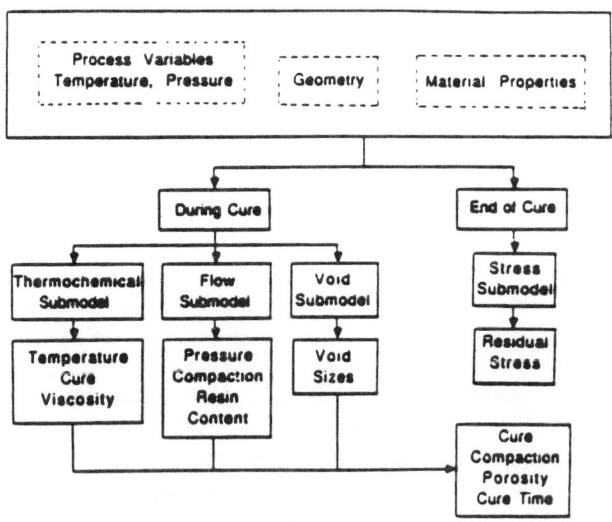

Figure 3.8 The results generated by the autoclave cure process simulation model (CURE code).

Table 3.5

Summary of the Input Parameters and the Results of the Autoclave Cure Model

Input Parameters
- Geometry
- Material properties
- Cure temperature and pressure as a function of time
- Bleeder (vacuum bag) pressure as a function of time

Results
- Thermo-Chemical Model
  - Temperature inside the composite as funcions of position and time
  - Degree of cure of the resin as functions of position and time
  - Resin viscosity as functions of position and time
- Flow Model
  - Pressure inside the composite as functions of position and time
  - Number of compacted prepreg plies as a function of time
  - Amount of resin in the bleeder as a function of time
  - Thickness and mass of the composite as a function of time
  - Cure time
- Void Model
  - Void sizes as functions of location and time
- Stress Model
  - Residual stresses and strains in each ply after cure

The major steps of the aforementioned calculation procedure are best illustrated by the following example.

1) The temperature distribution is calculated as a function of time (Figure 3.9). The temperature distribution should be uniform across the thickness of the composite and should not exceed a prescribed maximum value. Hence, the cure cycle resulting in temperatures shown in the left of Figure 3.9 is unacceptable. Different cure cycles must be tried until the temperature history inside the material is as shown in the right of Figure 3.9.
2) The degree of cure and viscosity distributions are plotted as a function of time (Figure 3.10). Since these parameters depend on temperature, and since the temperature distribution is uniform as shown in the right of Figure 3.9, the degree of cure and the viscosity will also vary uniformly across the material (Figure 3.10).

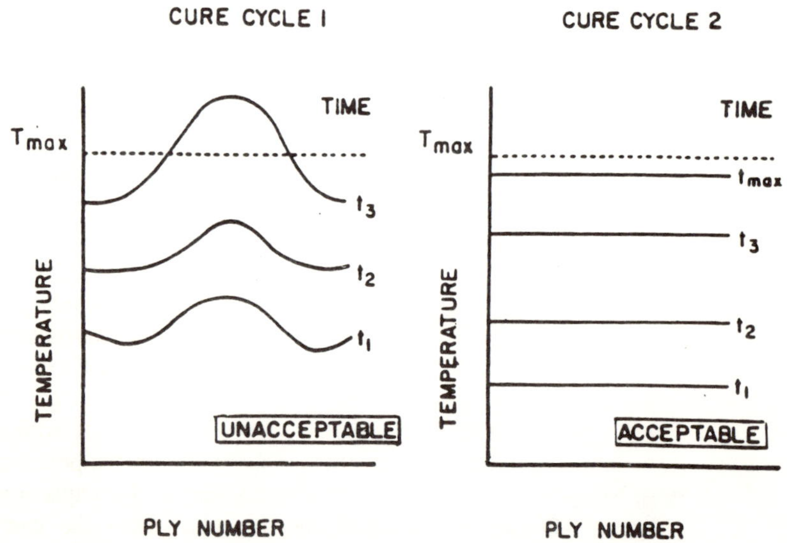

Figure 3.9 Typical temperature distributions across a laminate during autoclave cure. Temperatures on the left are unacceptable because they are nonuniform and because they exceed a prescribed maximum value.

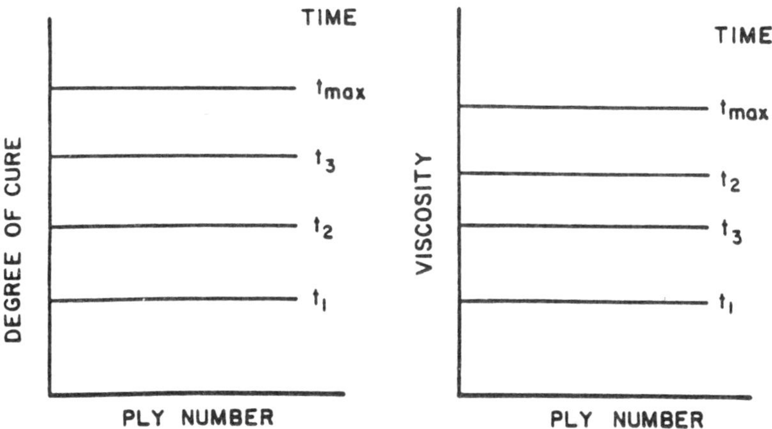

Figure 3.10 Typical degree of cure and viscosity distributions across a laminate during autoclave cure.

3) The variation of viscosity with time is plotted (Figure 3.11). From this plot the "cure window" (i.e., the viscosity region in which the material is still "soft") is established. If this cure window is too short or too long a new cure temperature is selected and the procedure, starting at step 1, is repeated.
4) The compaction as a function of time across the composite is plotted (Figure 3.12). At the end of the cure process every ply must be compacted. Hence, a process leading to the compaction in the left of Figure 3.12 is unacceptable. Another cure cycle must be chosen which ensures that compaction is complete across the entire laminate (Figure 3.12, right). The rate at which compaction progresses is governed both by the cure temperature and the cure pressure. Therefore, the speed at which compaction takes place can be increased by changing either the cure temperature, the cure pressure, or both. If the rate of compaction is unacceptable, a new cure cycle is selected and the iteration is repeated starting at step 1.
5) The degree of cure versus time is plotted (Figure 3.13). From this plot the cure time is determined. If this time is deemed too long, a new cure cycle is selected, and again the iteration is repeated starting at step 1.

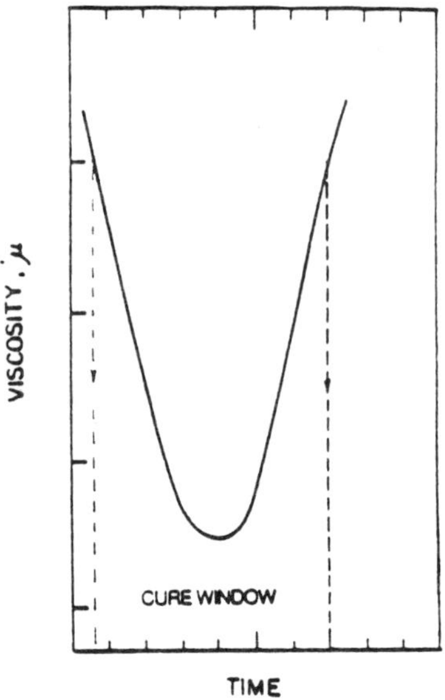

Figure 3.11 Viscosity as a function of time during autoclave cure, and illustration of the cure window.

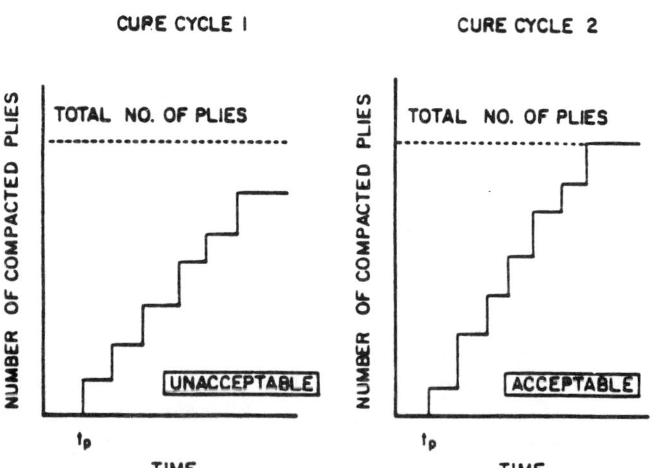

Figure 3.12 Compaction during autoclave cure of a laminate. Compaction on the left is unacceptable because not all the plies are compacted.

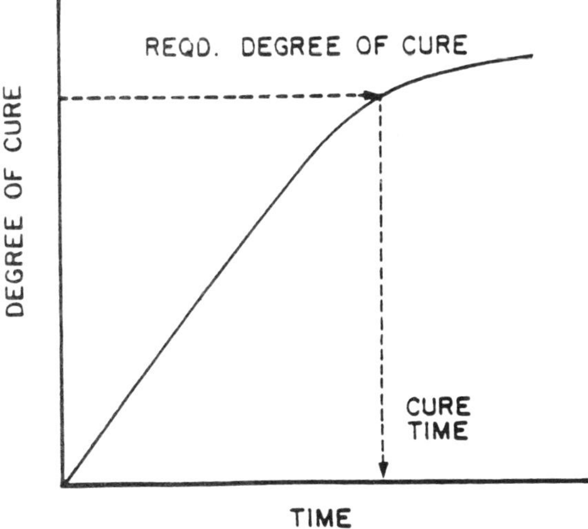

Figure 3.13 Degree of cure versus time during autoclave cure of a laminate.

In addition to the above steps, the cure cycle can further be optimized to provide the desired residual stresses and strains, and void content. The procedure leading to the optimum values of these parameters is much the same as that given for the optimization of the temperature distribution, cure window, compaction, and cure time.

CHAPTER 4

An Expert System for the Autoclave Cure Process

It is desired to cure inside an autoclave a part made of a fiber reinforced thermosetting matrix composite. The cure process must satisfy two major objectives:

1) The part must be of good quality, i.e. it must have the required properties;
2) The cure must be accomplished in the shortest time.

To achieve these objectives, the requirements listed in Table 2.1 must be met.

The problem at hand is to select and control in real time the cure conditions (autoclave temperature and pressure) so that all six requirements in Table 2.1 are satisfied. The model described in the previous chapter can be used to select the processing conditions but can not be used to control the conditions inside the autoclave. The cure conditions can be selected and controlled simultaneously and in real time by a rule based expert system.

Curing by an expert system requires sensors, rules, and a data processing algorithm. A major objective is to perform the expert system function (i.e. to control the cure) with the minimum number of sensors and with the simplest possible rules. In the system presented here, only the following parameters need to be monitored during cure (Figure 4.1)

1) the autoclave temperature,
2) the surface temperatures of the composite,
3) the midpoint temperature of the composite,
4) the dielectric properties at one point inside the composite,
5) the thickness of the composite, and,
6) the autoclave pressure.

Autoclave

Temperature and Pressure

● T_a ● P

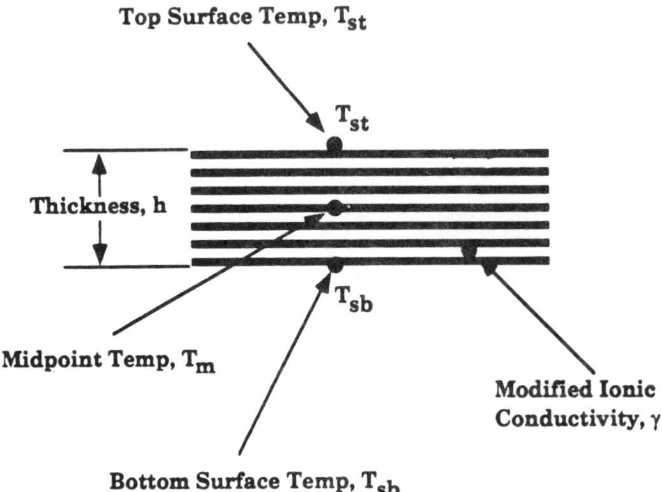

$$T_s = \frac{T_{st} + T_{sb}}{2} \qquad \Delta T = |T_m - T_s|$$

Figure 4.1 Illustration of the sensor locations in the composite.

Data provided by these sensors are input to the rules which, in turn, adjust - in real time - the autoclave controls (heater, cooler, pressure).

The procedure does not require prior information of the material properties or geometry. With the exception of the maximum permissible temperatures and pressure, no property need be specified. Furthermore, the system is not limited to any specific thermosetting matrix. Finally, no restrictions are placed on the shape or size of the part.

Two major expert system strategies have been proposed in recent years. According to one approach, pursued by researchers at McDonnell Douglas Corporation [31], the outputs of the sensors are combined with the results of analytic models to establish the cure conditions. This approach obviously requires models as well as prior information (data) regarding the material properties. According to another approach, proposed by researchers at the Air Force Materials Laboratory [17, 18] and at Stanford University [19], the sensor outputs are combined directly with the rules. Here, we present an expert system based on this approach.

The control strategy and the rules which satisfy the autoclave cure requirements listed in Table 2.1 can be formulated in different ways. The strategy adopted here was selected for its simplicity and its effectiveness. The first step in the strategy, introduced to meet the temperature requirements (points 1 and 2 in Table 2.1), is built on the concept proposed by Abrams and her coworkers [17], who were the first to employ an expert system to cure composite laminates. However, the present control strategy differs somewhat from that given in reference [17]. Rules pertaining to compaction, curing stresses and voids were not proposed by Abrams, et. al. and, to the best of our knowledge, were put forth for the first time by Ciriscioli [19].

The main steps of the overall control strategy are summarized below. For clarity, the various steps are discussed separately. In practical applications they must, of course, be implemented simultaneously.

The control strategy is devised on the premise that during the cure process only three controls can be adjusted:

1) the autoclave heaters,
2) the autoclave coolers and,
3) the autoclave pressure.

In some autoclaves active cooling cannot be applied before the composite is fully cured. In this case, the autoclave temperature decreases somewhat when the heaters are turned off due to inactive cooling. Even in such autoclaves, the autoclave can be cooled actively after the composite has been cured completely.

4.1 TEMPERATURE

During the curing process (i.e. before the composite is fully cured) the rules listed in Table 4.1 are invoked to ensure that a) the temperature does not exceed a preset maximum value T_{max}, b) the temperature distribution is fairly uniform, and c) no significant exotherm occurs inside the composite.

The maximum temperature T_{max} depends on the material properties and on the layup. Each material system has a maximum temperature which must not be exceeded. In addition, the temperature is limited by the layup, because some layups become damaged if cured at high temperatures (see Rule 5 and Section 6.5)

In Table 4.1, T_a is the autoclave temperature, T_m is the temperature at the midpoint of the composite, and T_s is the average surface temperature (Figure 4.1)

$$T_s \equiv \frac{T_{st} + T_{sb}}{2} \tag{4.1}$$

where T_{st} and T_{sb} are the temperatures at two opposite surfaces. The temperature difference ΔT is defined as

$$\Delta T \equiv |T_m - T_s| \tag{4.2}$$

ΔT^H_{max} is the maximum allowed difference between the composite midpoint and surface temperatures during heating.

The rules regarding the autoclave, surface, and midpoint temperatures are obvious. The rules regarding the rate of change of the midpoint temperature ($\ddot{T}_m = d^2T_m / dt^2$) are used to ensure that the temperature rise does not accelerate unduly at the midpoint, thereby minimizing the extent of the exotherm (Figure 4.2)

The reason for the second set of rules in Table 4.1 is as follows. Once an exotherm has occurred and has been brought under control, the average surface temperature may be less than the midpoint temperature ($T_s < T_m$) while, simultaneously, the midpoint temperature may be decreasing ($\ddot{T}_m < 0$). Under these conditions the difference between the average surface temperature and the midpoint temperature may become too large ($\Delta T > \Delta^H_{max}$). In this situation the temperature difference can only be reduced by heating the surface.

The initial heat up rate can somewhat be accelerated by observing that in the early stages of the cure (before the midpoint temperature becomes too high) there is no likelihood of exotherms. Hence, in this stage it is unnecessary

Table 4.1

Temperature Rules During Cure ($\dot{\gamma} \neq 0$)

RULE 1

IF

$T_a < T_{max}$ ***and*** $T_s < T_{max}$ ***and*** $T_m < T_{max}$ ***and*** $\Delta T < \Delta T^H_{max}$ ***and*** $\ddot{T}_m < 0$

THEN

Increase the autoclave air temperature at maximum rate
(i.e. turn on the autoclave heaters at maximum power)

IF

$T_a < T_{max}$ ***and*** $T_s < T_{max}$ ***and*** $T_m < T_{max}$ ***and*** $\ddot{T}_m < 0$ ***and***
$\Delta T > \Delta T^H_{max}$ ***and*** $T_s < T_m$ ***and*** $\dot{T}_m < 0$

THEN

Increase the autoclave air temperature at maximum rate
(i.e. turn on the autoclave heaters at maximum power)

IF

$T_a > T_{max}$ ***or*** $T_s > T_{max}$ ***or*** $T_m > T_{max}$ ***or*** $\Delta T > \Delta T^H_{max}$ ***or*** $\ddot{T}_m > 0$

THEN

Decrease the autoclave air temperature at maximum rate
(i.e. turn off the autoclave heaters and, if possible, turn on the coolers)

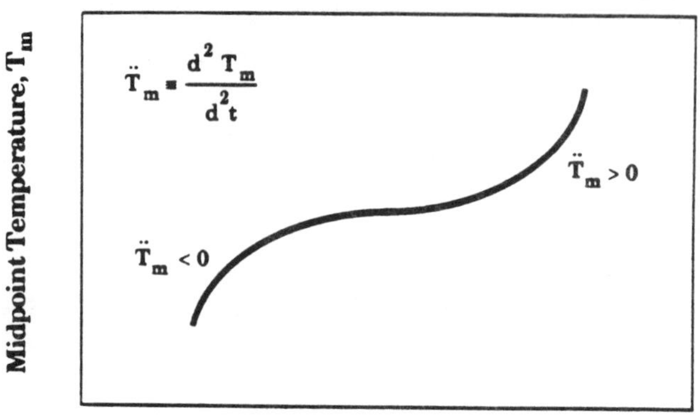

Figure 4.2 Illustration of the rate of change of laminate midpoint temperature.

to enforce the rules that $\Delta T < \Delta^{H}_{max}$ and $\ddot{T}_m < 0$. By ignoring these rules the initial heat up process is accelerated. The midpoint temperature below which these two rules are ignored must, of course, be specified if advantage is to be taken of these additional requirements on ΔT and $\ddot{T}_m$.

The aforementioned rules require the measurements of only four temperatures: in the autoclave, at the midpoint, and at the two surfaces on the composite. As will be shown subsequently, it is unnecessary to measure temperatures at additional points even for very thick (6 - 8 in, 1000 plies) laminates.

4.2 COMPACTION

During cure it must be ensured that all the plies are fully compacted before the material becomes so viscous that resin cannot flow under the applied pressure. Compaction is accompanied by a change (decrease) in thickness

$$\Delta h = h_o - h \tag{4.3}$$

where h_o is the original (time $t = 0$) thickness of the uncompacted composite and h is the thickness at time t.

Obviously, both the applied pressure and temperature play important roles in the compaction process. The pressure provides the force needed to squeeze the resin out, while the temperature affects the resin viscosity and hence the resin flow. In the present expert system it is stipulated that the maximum possible pressure is applied at the start of the cure (time $t = 0$). Contrary to some belief, there is no scientific evidence that such early pressure application affects the quality of the product. It is noted that, if desired, the pressure could be applied at a later time. Once the pressure is applied, the rate of compaction is controlled only by the autoclave temperature.

At the start of cure, heating is set at maximum rate to reduce rapidly the viscosity. For a period, from time $t = 0$ to $t = t_{c_o}$ (Figure 4.3), resin flow is

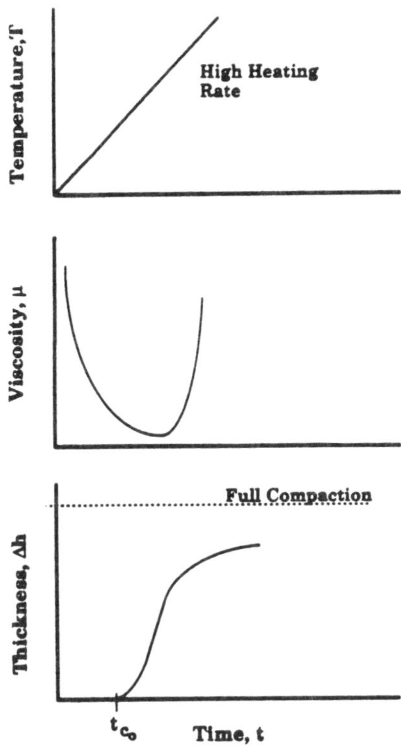

Figure 4.3 Illustration of the relationship between temperature, viscosity and thickness change (compaction) at a high heating rate.

negligible and compaction is practically zero. At time $t > t_{c_o}$ resin flow and compaction become significant. At this time, the heating rate could be maintained at a high level, reducing further the resin viscosity and increasing the compaction rate (Figure 4.3). However, the fast rise in temperature resulting from the high heating rate would result in a subsequent rapid rise in viscosity, as illustrated in this figure. In this case the time period in which the viscosity is sufficiently low may be too short for all the excess resin to flow out and for all the plies to compact, as illustrated in the bottom of Figure 4.3.

To prevent the rapid drop, followed by a rapid rise in viscosity, it is proposed here to keep the viscosity nearly constant from the start of the compaction $t = t_{c_o}$ until compaction is completed (Figures 4.4 and 4.5). The

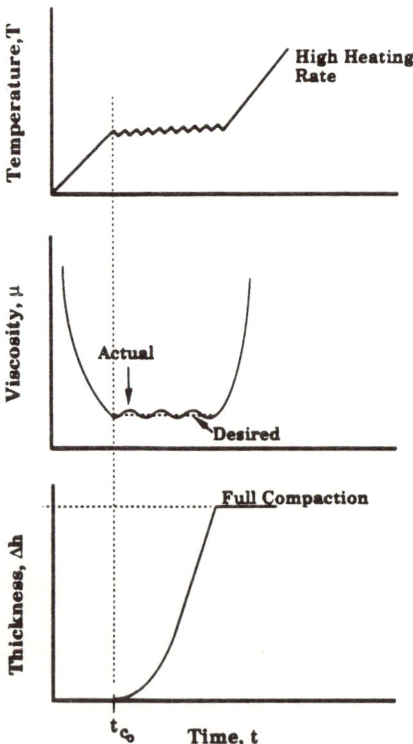

Figure 4.4 Illustration of the desired compaction control strategy.

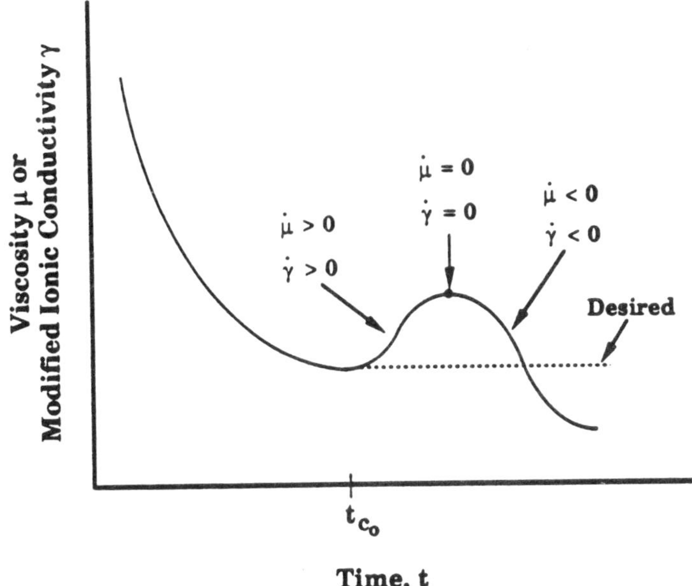

Figure 4.5 Illustration of the viscosity and modified ionic conductivity behavior resulting from the control strategy.

viscosity μ is maintained at the desired level by alternately raising and lowering the autoclave temperature. (At constant temperature, the resin viscosity would steadily increase). When the viscosity increases ($\dot{\mu} = d\mu / dt > 0$) the heaters are turned on until the viscosity starts decreasing ($\dot{\mu} = 0$, Figure 4.5). At this time, the heaters are turned off (and, if possible, the coolers are turned on) until either the viscosity starts increasing again or until the rate of compaction slows down (see below).

Unfortunately, at the present time no known method exists for measuring the viscosity during cure [32]. However, the ionic conductivity Σ, which is related to viscosity, can be measured. The rules require only the trends in the data and not the absolute values. Hence, the rules are formulated using the inverse ionic conductivity $1/\Sigma$, which exhibits the same trend with time as the viscosity. For convenience, the rules are formulated in terms of a modified ionic conductivity defined as

$$\gamma \equiv \frac{1}{\log_{10} \Sigma} \tag{4.4}$$

where Σ is the ionic conductivity [32]. The time derivative of γ is $\dot{\gamma} = d\gamma / dt$, illustrated in Figure 4.5.

It is further desired that the compaction proceed at a high rate. To prevent the "slowing down" of the compaction, the additional rule is imposed that heat is applied whenever the rate of compaction decreases (Figure 4.6), i.e. heat is applied whenever

$$\Delta\ddot{h} \equiv \frac{d^2(\Delta h)}{dt^2} < 0 \tag{4.5}$$

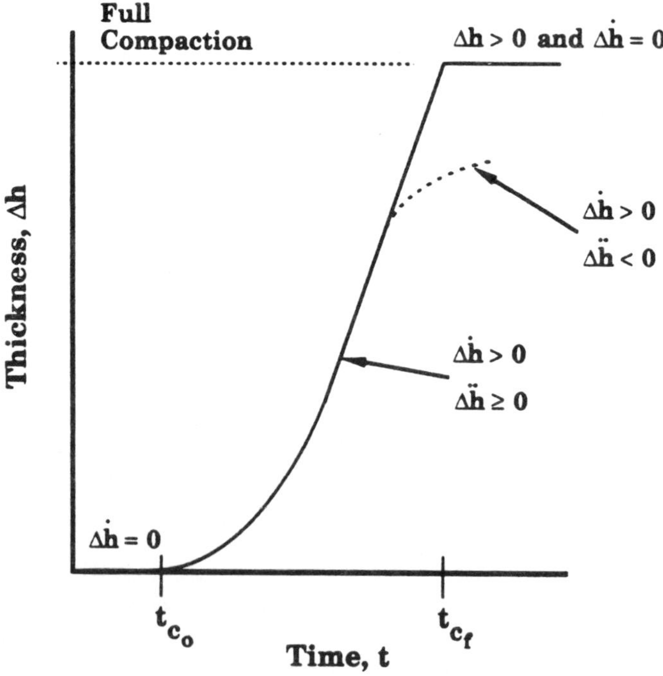

Figure 4.6 Illustration of the thickness behavior resulting from the control strategy.

Finally, the compaction is ended (and the composite cannot be compacted further) when $\Delta\dot{h} = (d\Delta h / dt)$ becomes zero. At this time ($t = t_{cf}$), the heating rate is increased to its maximum according to Rule 1.

The condition $\dot{h} = 0$ (when $\Delta h > 0$) does not insure that every ply has been compacted. Whether or not full compaction occurred may only be established if the initial excess volume fraction V_{bf} ("bulk factor") is known. Thus we have

$$\frac{\Delta h}{h_o}\begin{cases} = V_{bf} & \text{Full Compaction} \\ < V_{bf} & \text{Incomplete Compaction} \end{cases} \tag{4.6}$$

The rules combining the requirements regarding the modified ionic conductivity and the thickness (compaction) are summarized in Table 4.2.

Table 4.2

Compaction Rules During Cure ($\dot{\gamma} \neq 0$)

RULE 2

IF	$\Delta h = 0$ ***and*** $\Delta\dot{h} = 0$	THEN	Increase the autoclave air temperature at maximum rate (i.e. turn on the autoclave heaters at maximum power)
IF	$\Delta h > 0$ ***and***		
	$\Delta\ddot{h} < 0$ ***or*** $\dot{\gamma} > 0$	THEN	Increase the autoclave air temperature (i.e. turn on the autoclave heaters)
	$\Delta\ddot{h} \geq 0$ ***or*** $\dot{\gamma} \leq 0$	THEN	Decrease the autoclave air temperature (i.e. turn off the autoclave heaters and, if necessary, turn on the coolers)
IF	$\Delta h > 0$ ***and*** $\Delta\dot{h} = 0$	THEN	Increase the autoclave air temperature at maximum rate (i.e. turn on the autoclave heaters at maximum power)

The rules in Table 4.2 require that the thickness and the ionic conductivity be measured. The thickness should be measured at the thickest portion of the composite. The ionic conductivity should be measured at the corresponding location, between two plies where the compaction is expected to be the slowest.

It is emphasized that, for some geometries, it may be impossible to compact all the plies under a given pressure. In this case, the expert system will lead to a composite which is highly compacted (likely to a higher degree than a composite cured by conventional methods) but is not fully compacted. Under these circumstances composites must be fabricated by alternate techniques, for example by debulking some ply segments separately.

4.3 VOIDS

The void growth may be suppressed and the void content may be minimized by applying a pressure which is higher than the saturation pressure P_s of the volatiles in the resin. Accordingly, the rule is imposed that the autoclave pressure must exceed the highest saturation pressure in the composite (see Table 4.3). The value of the saturation pressure is determined by either the midpoint or the average surface temperature (T_m or T_s Figure 4.1), whichever is higher. The values of P_s, corresponding to given temperatures, are determined with the use of property tables.

Table 4.3

Void Rules During Cure ($\dot{\gamma} \neq 0$)

RULE 3

IF

$P < P_s$ ***and*** $P < P_{max}$

THEN

Increase the autoclave pressure

IF

$P > P_s$ ***or*** $P > P_{max}$

THEN

Decrease the autoclave pressure

Since the temperature changes with time, so does the saturation pressure P_s and the required autoclave pressure P. The pressure can be adjusted continuously (subject to Rule 3) or applied at the start of the cure (at $t = 0$), but must be below a preset permissible value P_{max}.

4.4 END OF CURE

The cure is considered to be complete when the rate of change of the degree of cure $\dot{\alpha}$ becomes zero ($\dot{\alpha} = d\alpha / dt$). Currently, there is no instrument available for directly measuring the degree of cure α [32]. However, the modified ionic conductivity (a parameter which can be measured) provides a qualitative measure of α. For convenience, the rules again are expressed in terms of γ (Eq. 4.4). Specifically, the cure is taken to be complete when $\dot{\gamma}$ becomes zero ($\dot{\gamma} = d\gamma / dt = 0$). The corresponding rule is stated in Table 4.4. At the time the cure becomes complete (i.e. when $\dot{\gamma} = 0$), the coolers are turned on to reduce the autoclave temperature.

Table 4.4

End of Cure Rules

RULE 4

IF	$\dot{\gamma} \neq 0$	**THEN**	Continue the cure
IF	$\dot{\gamma} = 0$	**THEN**	Decrease the autoclave air temperature (i.e. turn off the autoclave heaters and, if possible, turn on the coolers)

4.5 RESIDUAL (CURING) STRESS

During cool down (i.e. after the cure has been completed) residual (curing) stresses are introduced into composites. As is shown in Chapter 6, for a given material and for a given layup the residual stresses depend mostly on the maximum temperature of the composite during cure or the post cure. The residual stresses seem to be unaffected by the cooling rate or the degree of cure (at least when α is sufficiently high, $\alpha > 0.7$ say), but are very sensitive

Table 4.5

Residual (Curing) Stress Rules

RULE 5

IF

(The layup is subject to damage by residual stresses)

THEN

$$T_{max} \leq T_{crit}$$

---

to layup. Thus, to avoid damage introduced by excessive residual stresses the composite's temperature must be kept below a preset value $T_{max} < T_{crit}$. The corresponding rule is given in Table 4.5 The appropriate value of T_{crit} must be determined by test for a particular layup and material.

Note that full cure (i.e. $\alpha \rightarrow 1$) may not be achieved at temperatures below T_{crit}.

4.6 END OF MANUFACTURING PROCESS

The manufacturing process is ended when (during cool down) the autoclave temperature T_a reaches room temperature T_{rt}. The appropriate rules are listed in Table 4.6.

Table 4.6

End of Manufcturing Process Rules ($\dot{\gamma} = 0$)

RULE 6

IF $T_a > T_{rt}$	**THEN**	Decrease the autoclave air temperature (i.e. turn off the autoclave heaters and, if possible, turn on the coolers)
IF $T_a \leq T_{rt}$	**THEN**	Stop process - open autoclave

4.7 TEMPERATURE CONTROL STRATEGY SUMMARY

In the foregoing, the various steps of the control strategy were stated separately. In practice, all these steps must be applied simultaneously. This introduces an added complexity since some of the rules pertaining to the autoclave temperature may be contradictory. In such situations, the dominant rule must be selected and applied. The overall temperature control strategy is summarized in Tables 4.7 and 4.8 as an example of the complexity required.

4.8 IMPLEMENTATION

For practical control of an autoclave, the expert system rules given in tables 4.1 to 4.6 must interface with the sensor outputs on the one hand, and on the other hand, with the controller inputs (Figure 4.7). These interfaces and the rules were incorporated in an algorithm designated as SECURE. The SECURE code was written in C language and, presently, is installed on a Macintosh II conputer. However, the code is sufficiently general, so that it could be transferred to other computers with only minimum modifications.

The inputs to the code are a) the preset values of some parameters and b) the outputs (signals) from the sensors. The preset values are the permissible maximum temperature T_{max}, the permissible maximum temperature differences ΔT^H_{max}, the permissible maximum pressure P_{max}, and the saturation pressure P_s (P_{max} and P_s are not needed for the operation of the code.)

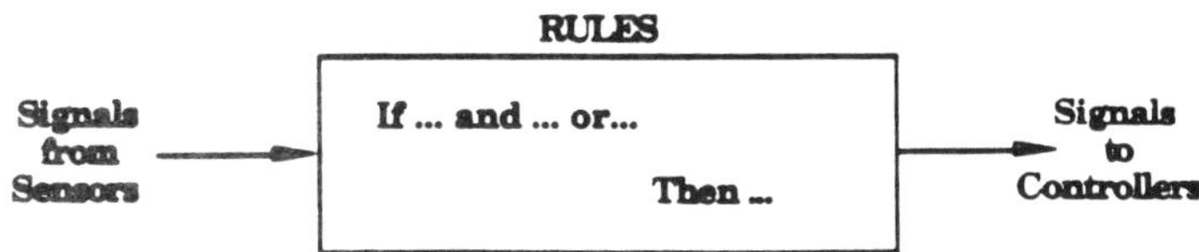

Figure 4.7 Required interfacing scheme for the SECURE code.

Table 4.7
Temperature Rule Summary (Heating)

TEMPERATURE RULE SUMMARY - HEATING

IF

$$\dot{\gamma} \neq 0 \ \textit{and}$$
$$T_a < T_{max} \ \textit{and} \ T_s < T_{max} \ \textit{and} \ T_m < T_{max} \ \textit{and} \ \Delta T < \Delta T_{max}^{H} \ \textit{and} \ \ddot{T}_m < 0$$

OR

$$\dot{\gamma} \neq 0 \ \textit{and}$$
$$T_a < T_{max} \ \textit{and} \ T_s < T_{max} \ \textit{and} \ T_m < T_{max} \ \textit{and} \ \ddot{T}_m < 0 \ \textit{and}$$
$$\Delta T > \Delta T_{max}^{H} \ \textit{and} \ T_s < T_m \ \textit{and} \ \dot{T}_m < 0$$

OR

$$\dot{\gamma} \neq 0 \ \textit{and} \ \Delta h = 0 \ \textit{and} \ \Delta \dot{h} = 0$$

OR

$$\dot{\gamma} \neq 0 \ \textit{and} \ \Delta h > 0 \ \textit{and} \left(\Delta \ddot{h} < 0 \ \textit{or} \ \dot{\gamma} > 0 \right)$$

OR

$$\dot{\gamma} \neq 0 \ \textit{and} \ \Delta h > 0 \ \textit{and} \ \Delta \dot{h} = 0$$

THEN

Increase the autoclave air temperature
(i.e. turn on the autoclave heaters)

Table 4.8
Temperature Rule Summary (Cooling)

TEMPERATURE RULE SUMMARY - COOLING

IF

$\dot{\gamma} \neq 0$ ***and***

$\left(T_a > T_{max} \text{ or } T_s > T_{max} \text{ or } T_m > T_{max} \text{ or } \Delta T > \Delta T^H_{max} \text{ or } \ddot{T}_m > 0\right)$

OR

$\dot{\gamma} \neq 0$ ***and*** $\Delta h > 0$ ***and*** $\Delta \ddot{h} \geq 0$ ***and*** $\dot{\gamma} \leq 0$

THEN

Decrease the autoclave air temperature
(i.e. turn off the autoclave heaters and, if possible, turn on the coolers)

The sensor signals which are utilized in the current version of the SECURE code are (Figure 3.1): the autoclave temperature T_a, the composite top surface, bottom surface and midpoint temperatures T_{st}, T_{sb}, T_m, the composite thickness change Δh, and the modified ionic conductivity γ. The outputs of these sensors are analyzed by the code. It would be convenient to have a separate computer handling the outputs of each of these sensors and the corresponding rules. This however is uneconomical. Therefore, an effort was made to implement the rules by numerical procedures which require only a single small microcomputer. Since such a computer can only handle one type of data at a time, the sampling intervals must be adjusted to accommodate the sequential nature of the data collection and analysis.

The fact that SECURE requires only a small microcomputer is a very important feature. The expert system previously developed by Abrams et. al. [17] and the expert system now being developed by McDonnell Douglas [31] require the simultaneous use of two or more central processing units.

In the following, the procedures are described which are used to analyze the data. It is emphasized that the algorithm proposed here is not the only one suitable for the task. It was adopted because it was found to be practical and efficient.

Temperature

The outputs of four temperature sensors are recorded: the autoclave temperature T_a, the composite top surface temperature T_{st}, the composite midpoint temperature T_m, and the composite bottom surface temperature T_{sb} (Figure 4.1). Fifty readings of each are taken, 1 μs apart. These values are averaged to give the value of each of the four temperatures over the 0.1 second interval of the readings.

This procedure is repeated every 2 seconds for 18 seconds, giving 10 data points. The average surface temperature T_s (Eq. 4.1) and the temperature difference ΔT (Eq. 4.2) are evaluated from the last set of data i.e. from the 10th data point. These values of T_s and ΔT, along with the autoclave temperature T_a and the midpoint temperature T_m corresponding to the last data, are used in the rules.

A third order polynomial in time is fitted to the midpoint temperature data. Differentiation of this polynomial with respect to time gives the first and second time derivatives of the midpoint temperature $\dot{T}_m$ and $\ddot{T}_m$. The values of these derivatives are evaluated at the time of the last (i.e. 10th) data point.

The above procedure is repeated every 50 seconds. The intervening 50 second interval is used to determine the ionic conductivity.

Compaction

The change in thickness Δh (compaction) is recorded 50 times, 0.002 seconds apart. These values are averaged to give the value of Δh over the 0.1 second interval of the readings. This procedure is repeated after a lapse of 20 seconds, then after a lapse of 50 seconds, then again after a lapse of 20 seconds, and so on until 10 data points are collected over a five minute period. This staggering of the data collection is needed to free up the computer for sampling of the temperature and ionic conductivity data.

A third order polynomial in time is fitted to the Δh data. Differentiation of this polynomial with respect to time gives the first and second time derivatives of the compaction $\Delta\dot{h}$ and $\Delta\ddot{h}$. The values of these derivatives are evaluated at the time of the last (i.e. 10th) data point.

Ionic Conductivity

Measurement of the modified ionic conductivity γ requires the longest sampling time, at least for the Micromet Eumetrics II instrument used in this study. This instrument directly provides the permittivity E' and the loss factor E''. The measurements must be performed at a frequency f for which the following conditions are satisfied [32]

$$\begin{aligned} 1 < E' < 25 \\ 12 << E'' << 1000 \end{aligned} \tag{4.7}$$

If the above conditions are met, then the ionic conductivity is calculated by the expression

$$\Sigma = 2\pi f E_o E'' \tag{4.8}$$

where E_o is the permittivity of free space ($E_o = 8.854 \times 10^{-12}$ Farads/meter). If the above conditions are not met, the frequency is changed and the procedure is repeated until the conditions in Eq. 4.7 are satisfied. Typically, it takes 50 seconds to arrive at an acceptable ionic conductivity value Σ.

The ionic conductivity measurements are repeated 10 times with a 20 second interval between measurements. (The 20 second intervals are used to perform analyses of other data.) From the measured value of ionic conductivity Σ, the modified ionic conductivity γ is calculated by Eq. 4.4. A second order polynomial is fitted to the γ versus time data. Differentiation of this polynomial with respect to time gives the first time derivative $\dot{\gamma}$. The value of the derivative is evaluated at the time of the last (i.e. 10th) data point.

In the present numerical scheme five frequencies (1, 10, 100, 1000, and 10000 Hz) are tried each time. If none of the data obtained by any of these five frequencies satisfies the requirements of Eq. 4.7, the time derivative $\dot{\gamma}$ of the modified ionic conductivity at that time step is not evaluated, but is approximated by its value at the previous time step.

Pressure

The autoclave pressure is recorded and is compared directly with the preestablished saturation pressure. The autoclave pressure may be set at the highest desired value at the start of the cure. The pressure may also be sampled and adjusted during the cure. In most cases sampling at 5 minute intervals should be adequate.

Control

The SECURE code determines and provides the appropriate control signals to the

1) pressure regulators,
2) heater controllers,
3) cooler controllers,

Although the SECURE code automatically controls the manufacturing process, for the operator's convenience the SECURE code displays the following information: autoclave temperature and pressure; composite surface and midpoint temperatures; time derivatives of the midpoint temperature, thickness change, and modified ionic conductivity; and the decisions made by the SECURE code. A sample display is given in Figure 4.8.

```
time 47.0 minutes
autoclave air temp (F) =185.4   autoclave pressure (psig) = 200.0
Tsurface (F) =183.2              Tmiddle (F) = 183.2
dT/dt = 6.841e-004               d2T/dt2 = -1.311e-004
dh/dt = 2.560e-005               dgamma/dt = 1.011e+000
Thermochemical event = Normal Reaction
Laminate is COMPACTED
decision is HEAT
```

Figure 4.8 Illustration of the SECURE code output on the screen during the autoclave control process.

CHAPTER 5

Verification of SECURE by Computer Simulation

In order to verify the overall control strategy presented in the previous chapter, the rules, and the resulting SECURE code must be validated. Validation was effected by employing the SECURE code in simulated as well as in actual cures. In this chapter, computer simulations are presented in which the SECURE code was used in conjunction with the CURE computer model (Chapter 3). In the simulations (and the actual cure experiments presented in the following chapter) flat laminates were simulated (or cured) in an autoclave controlled by SECURE. The laminates tested were made of either Fiberite T300/976 (HYE-3076E) tape, Fiberite T300/976 (HMF349D/76) woven fabric, or Hercules AS4/3501-6 tape, and ranged in thickness from 0.1 to 6.50 inches. Detailed geometries of these laminates are given in Table 5.1. The reason for using laminates with widely different thicknesses is obvious. The Fiberite and Hercules prepregs were used because they have very different cure kinetics (Table 3.4).

Simulations were accomplished by simulating the manufacturing process by the CURE computer model . To this end, the SECURE code was coupled to the CURE code. The combination of the CURE and SECURE codes was designated as SIMULATOR and was installed on a Macintosh II computer. In the operation of SIMULATOR, the outputs of the CURE code were used as inputs to the SECURE code; the outputs of the SECURE code were used as inputs to the CURE code, as illustrated in Figure 5.1.

The CURE code provides directly the viscosity μ and the degree of cure α, but not the modified ionic conductivity γ. Since only the time derivatives (and not the absolute values) of these parameters are of interest, during cure the time derivative of μ (μ being provided by CURE) was used in place of the time derivative $\dot{\gamma}$. The end of the cure was determined by the time

Table 5.1

Geometries of the Laminates Used to Validate the SECURE Code

Material	No. Plies	Cured Thickness (in)	Width(in)	Length (in)
Fiberite T300/976	16	0.104	10	12
(HYE-3076E)	52	0.338	10	12
6K Tape	200	1.375	3	12
	1000	6.500	3	12
Fiberite T300/976 HMF 349D/76 6K Woven Fabric	26	0.423	10	12
Hercules AS4/3501-6 (6K Tape)	52	0.393	10	12

Note:
For unidirectional tape the fibers are along the length (0° plies). For the fabric the fibers are in the length and width directions (0° and 90°). The exceptions are the 16 plies at the bottom surface (tape, or the 8 plies fabric) which are $[0_2/\pm45_2/90_2]_s$.

derivative of α ($\dot{\alpha} = d\alpha / dt = 0$). This is acceptable since, as has been shown previously [32], $\dot{\alpha}$ and $\dot{\gamma}$ become zero at about the same time.

The strain and void rules were not exercised in the SIMULATOR because the CURE code does not provide the needed information. The pressure (200 psig) was "applied" at the start of the manufacturing process.

With the exception of the 1000 ply laminate, all laminates listed in Table 5.1 were "cured" by the SIMULATOR to assess the operation of SECURE. For simplicity, the simulated cure assembly consisted only of a 0.375 inch aluminum tool plate and the composite, and a bleeder which had "infinite"

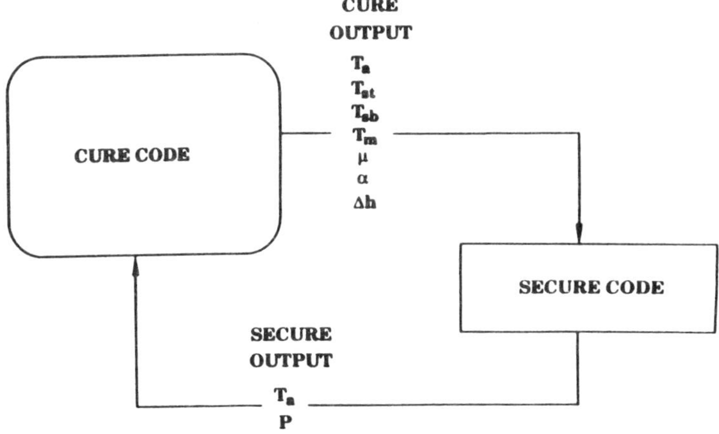

Figure 5.1 The SIMULATOR combining the CURE and SECURE codes.

porosity and thermal conductivity. The simulations were performed with the preset conditions that the temperature T_{max} must not exceed 390° F and the maximum temperature difference ΔT^H_{max} must be less than 12° F.

Typical results for the simulated cures of a thin (16 plys) and a thick (200 plys) Fiberite T300/976 unidirectional composite are shown in Figures 5.2 through 5.5. As is seen from these figures, SECURE performed well all its required tasks. It kept the maximum temperature below the allowed limit, and ensured that all the plies were compacted. The temperature difference on occasion exceeded the preset maximum value (12° F) Similar results were obtained for all the other laminates (Appendix C). From these results it was concluded that SECURE operated well and could be expected to perform satisfactorily in an actual autoclave environment.

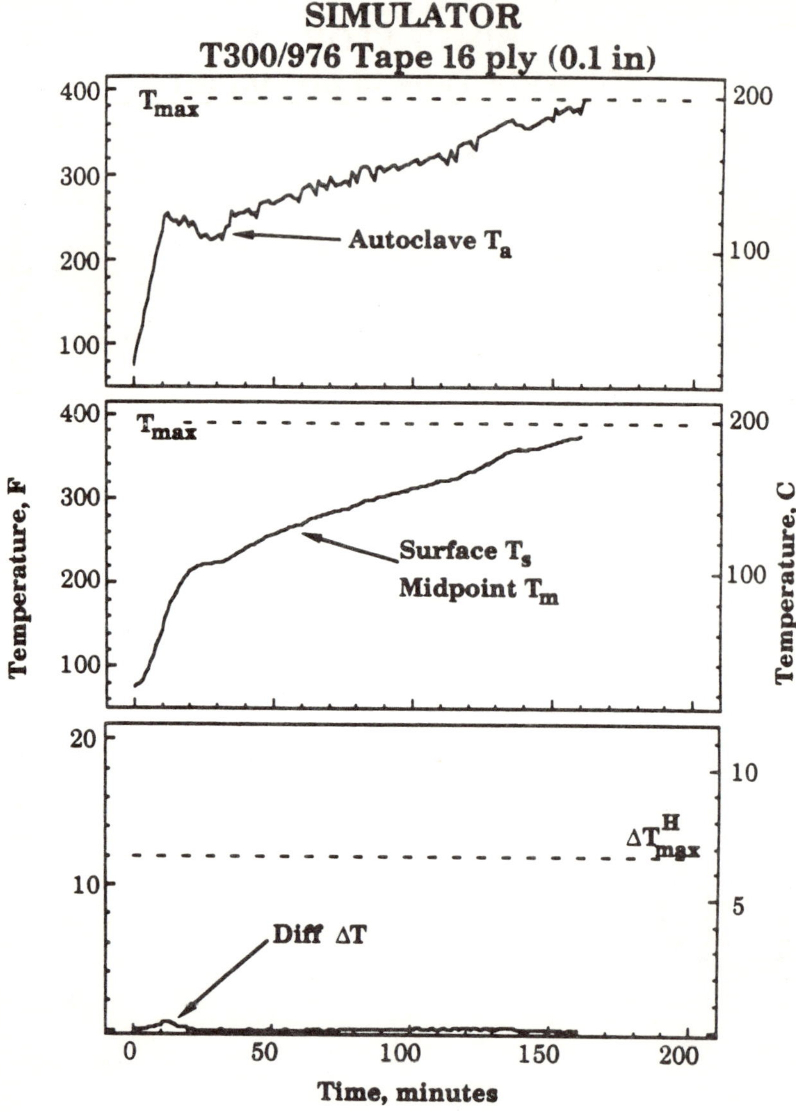

Figure 5.2 Temperature from the SIMULATOR (Table 5.1).

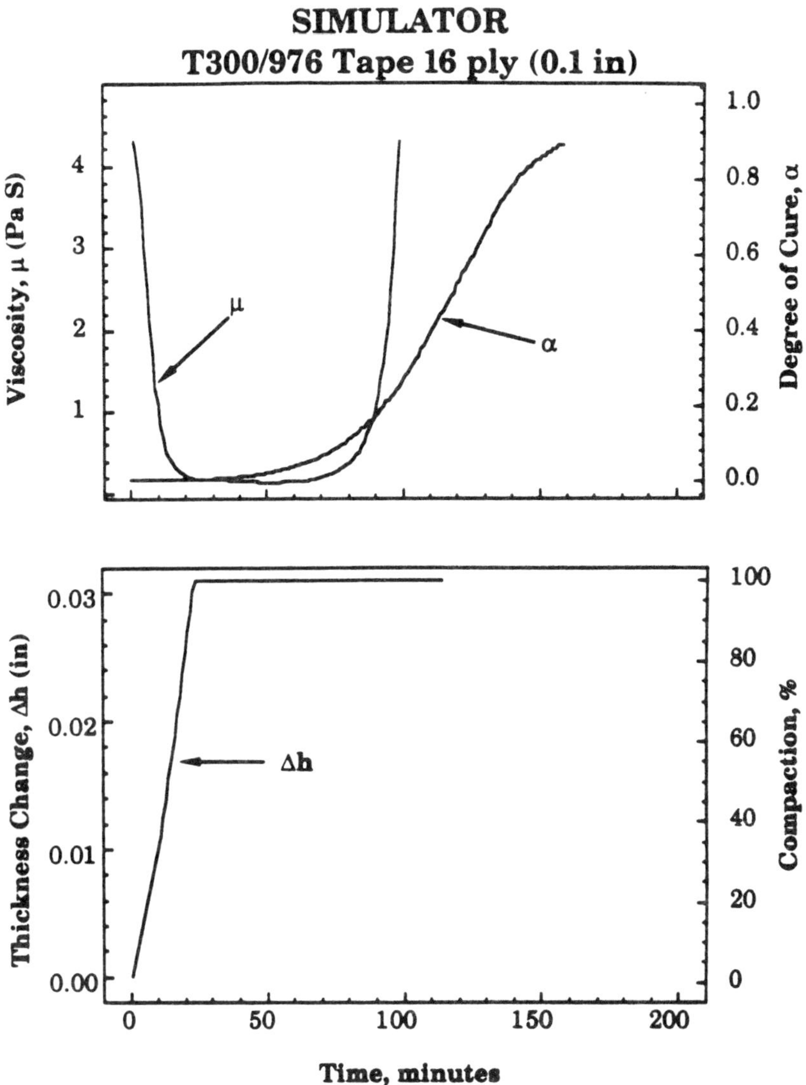

Figure 5.3 Viscosity, degree of cure and compaction from the SIMULATOR (Table 5.1).

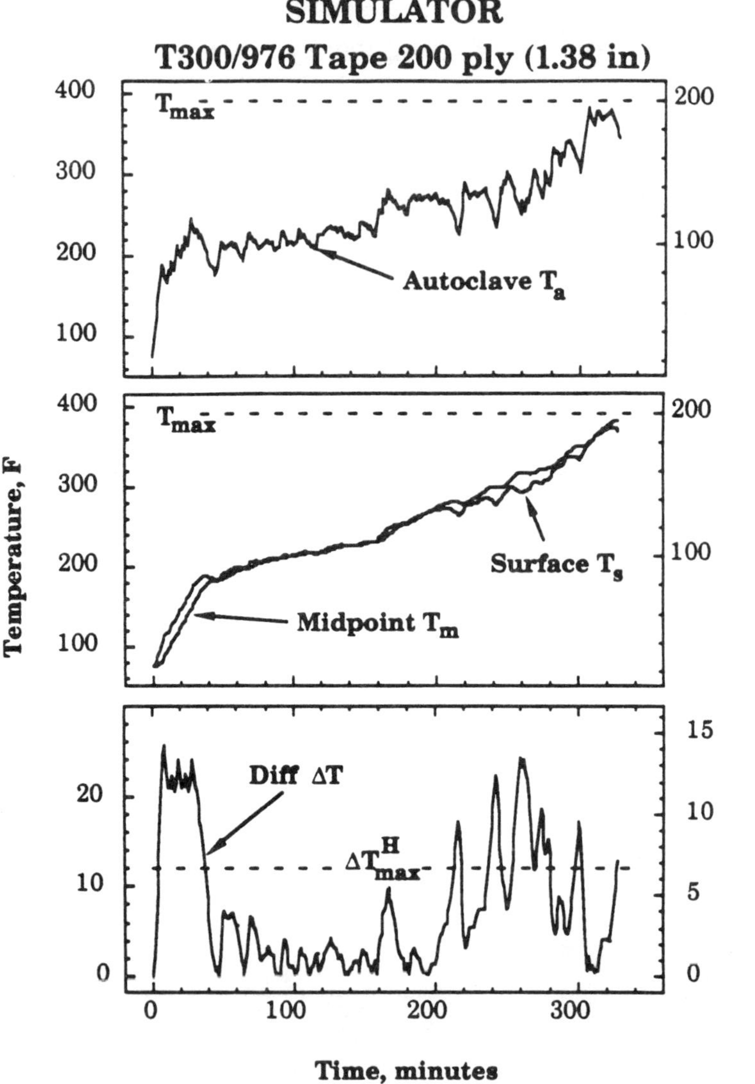

Figure 5.4 Temperature from the SIMULATOR (Table 5.1).

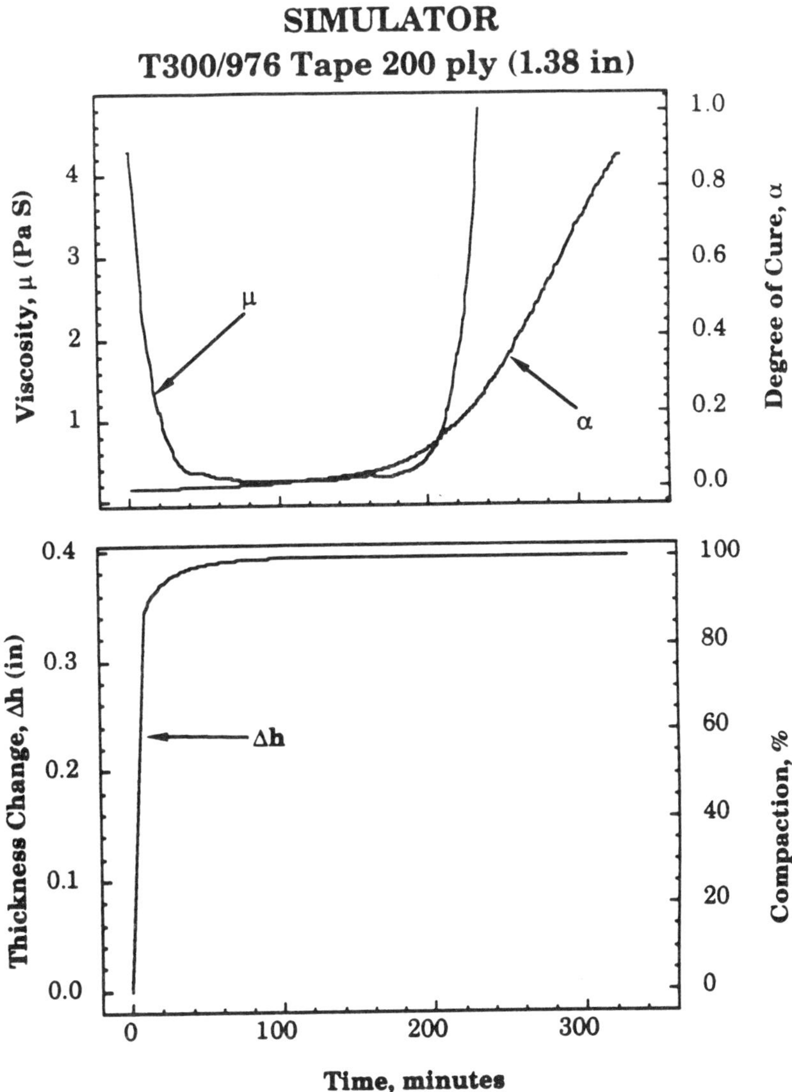

Figure 5.5 Viscosity, degree of cure and compaction from the SIMULATOR (Table 5.1).

CHAPTER 6

Verification of SECURE by Experiment

Experiments were conducted to evaluate the performance of SECURE and to assess the quality of composites fabricated by SECURE. Specifically, the tests were designed to determine how well SECURE performs its required functions (i.e. how well it controls the autoclave), and to establish what improvements in cure time and in mechanical properties are offered by SECURE. A total of eleven laminates were cured, six by SECURE and, for comparison, five by conventional cure cycles. The laminates were made of either Fiberite T300/976 tape or fabric, or Hercules AS4/3501-6 tape, and ranged in thickness from 16 to 1000 plies (Table 5.1). Five of the laminates cured by SECURE were identical to the five laminates cured by conventional methods. The 1000 ply Fiberite T300/976 laminate was cured only by SECURE.

6.1 EXPERIMENTAL APPARATUS AND PROCEDURE

Apparatus

The cure assemblies used for both the SECURE and the conventional cures are shown in Figure 6.1. Each cure assembly consisted of the following components

1) a 0.375 in. thick aluminum tool plate,
2) 1 layer of peel ply,
3) the composite,
4) 1 layer of porous teflon coated glass fabric,
5) bleeder plies, sufficient numbers to absorb all excess resin
6) 1 layer of nonporous film,

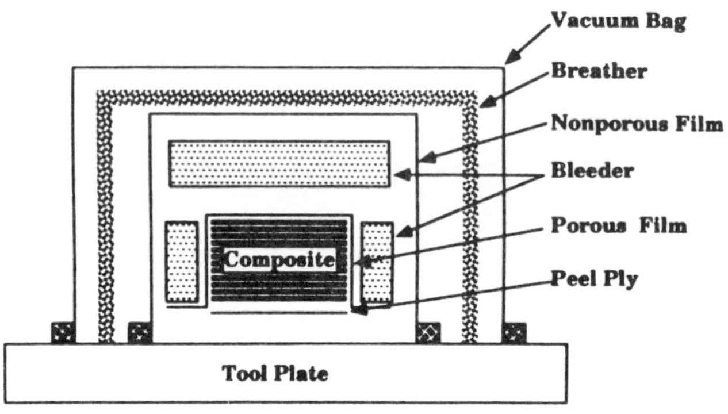

Figure 6.1 Components of the cure assembly.

7) 1 breather layer, and,
8) a vacuum bag.

Note that bleeders were placed around the four edges as well as on top of the laminate. Each assembly was instrumented by four thermocouples, a dielectric sensor, a thickness gauge, and a pressure gauge (Figure 4.1). Iron-Constantan thermocouples were placed in the autoclave (about 6 inches from the top of the vacuum bag), on the top and bottom surfaces of the laminate, and at the midpoint of the laminate. A Micromet Instruments Corporation Low Conductivity dielectric sensor was mounted between the bottom and next to bottom plies. This location was chosen because here the degree of cure was usually the lowest. A specially constructed thickness gauge was attached to the top surface of the laminate. Details of this sensor are given in Appendix D. Finally the pressures in the autoclave and inside the vacuum bag were monitored.

The signals from the sensors were fed into specially constructed signal conditioning circuitry (Appendix E) and thence to a data acquisition board installed in a Macintosh II microcomputer. The data were processed (Section 4.8) and stored in this computer.

Procedure

The cures were performed in one of two ways, either by a conventional method or by SECURE. In the conventional cures the autoclave temperature and pressure were set manually according to a predetermined schedule using the autoclave's temperature and pressure controllers. When curing by SECURE the autoclave temperature was controlled directly by SECURE.

During every cure (either by SECURE or by the conventional method) the autoclave temperature, composite midpoint and upper and lower surface temperatures, the composite thickness, and the modified ionic conductivity were measured (Figure 4.1).

The pressure was always set at the start of the cure at P = 200 psi. The one exception was the 16 ply T300/976 laminate cured by SECURE for which the pressure was 150 psi. When curing with SECURE the maximum temperature and the maximum temperature difference were specified as

$$T_{max} = 390^{\circ}\text{F} \qquad \Delta T^{H}_{max} = 12^{\circ}\text{F} \tag{6.1}$$

When curing with conventional cure cycles, autoclave temperatures were used which were determined by the CURE computer code as follows. For each laminate the pressure P = 200 psi and an assumed autoclave temperature history were entered into the CURE code. Then the temperatures inside the cure assembly, the degree of cure, and the compaction were calculated. The procedure was repeated with many different assumed temperature histories until an autoclave temperature was found which resulted in a) the required cure assembly temperatures ($T_{max} < 390^{\circ}\text{F}$), b) full compaction and c) the shortest cure time. The autoclave temperatures determined by this "optimization" are shown in Figure 6.2.

The output of SECURE was converted to electrical signals by the computer and these signals were used to adjust the autoclave heater control. The autoclave pressure controller was set manually, at the start of the cure, and the autoclave coolers were never activated. After the cure was completed, the heaters were turned off and the autoclave was cooled only by natural convection from the surrounding room air.

The laminates listed in Table 4.1 did not show signs of any damage. Therefore, to evaluate the residual stress rule, tests were also performed with laminates made of Fiberite T300/976 tape arranged as $[0_4/90_4]_s$. A total of five laminates were tested by curing and postcuring at different temperatures (Section 6.5). After both the cure and the postcure the laminates were X-rayed and inspected for damage.

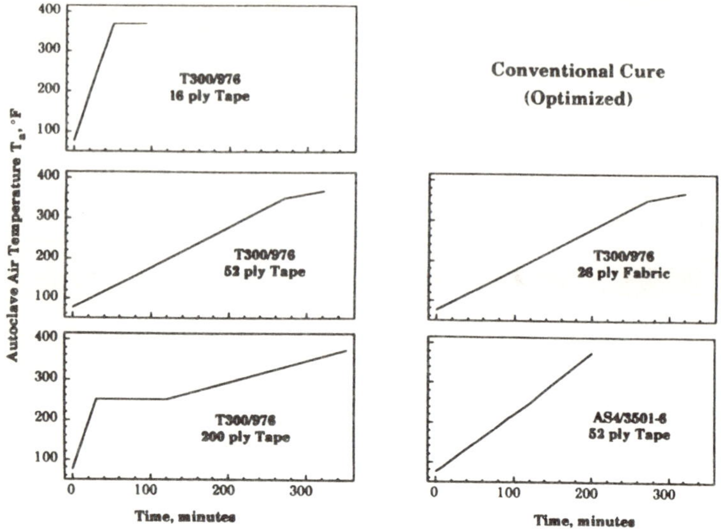

Figure 6.2 The autoclave air temperatures used in the optimized conventional cycles.

Evaluation

To verify the quality of the cured laminates, mechanical tests were performed on specimens taken from each laminate. Since many of the laminates were too thick for standard mechanical tests, special arrangements were needed to make test specimens of appropriate thicknesses. A porous teflon coated glass fabric (200TFP, Richmond Corporation) was placed between the 16th and 17th plies for tape and the 8th and 9th plies for cloth (the first ply being nearest to the tool plate). A 16 ply sublaminate was created in this manner which could readily be separated from the main laminate after the cure was completed (Figure 6.3). The layup in this sublaminate was $[0_2/45_2/90_2]_s$. Note that this layup is different than the unidirectional or crossply layup of the main laminate. The test specimens were then cut out of the $[0_2/45_2/90_2]_s$ sublaminate. (The 0° plies being along the longitudinal axis of each specimen.) The plies nearest to the tool plate were used to make the test specimens because these plies were expected to have the poorest properties since they were the least compacted, had the highest void content, and had the lowest degree of cure.

The tensile and compressive strengths and modulii, the short beam shear strengths, the fiber volumes, and the void contents were measured. The

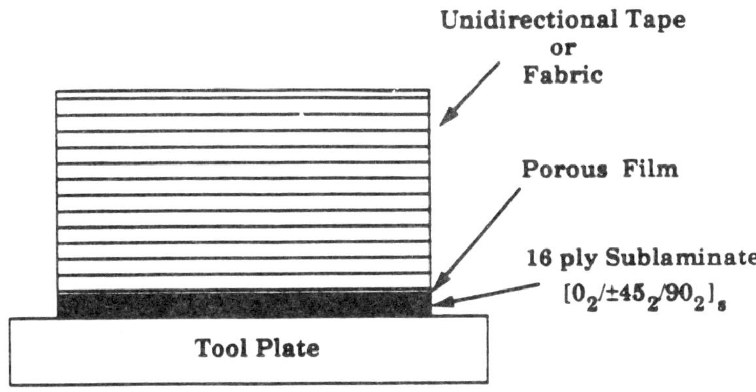

Figure 6.3 Illustration of the position and layup of the sublaminate used for the mechanical properties specimens.

geometries of the test specimens are shown in Figure 6.4. Fiberglass tabs were attached to the tensile and compressive specimens.

The compression tests were preformed by ASTM D3410 "Standard Test Method for Compressive Properties of Unidirectional or Crossply Fiber-Resin Composites", the tensile tests were performed by ASTM D3039 "Standard Test Method for Tensile Properties of Fiber-Resin Composites", and the short beam shear tests were performed according to ASTM D2344 "Standard Test Method for Apparent Interlaminar Shear Strength of Parallel Fiber Composites by the Short Beam Method". The tensile and compression tests were performed at a crosshead speed of 0.02 in/min. The short beam shear tests were performed at a crosshead speed of 1.0 in/min. In the tensile and compression tests, the strains were measured by a SD-101 extensometer (Measurements Technology Inc., Roswell GA). The fiber volumes and void contents were determined by ASTM D3171 "Standard Test Method for Fiber Content of Resin Matrix Composites by Matrix Digestion". All tests were conducted at room temperature.

The amount of compaction was also measured. Although this information is not necessary for the operation of SECURE, it was obtained to determine if SECURE indeed resulted in fully compacted laminates. The amount of compaction was evaluated by measuring the thickness of the 16 ply T300/976 laminate before and after cure. The compaction per ply was found to be 0.002 in. This value was used for the T300/976 tape laminates (twice this value was used for the fabric laminate) as well as for the Hercules AS4/3501-6 laminates [2].

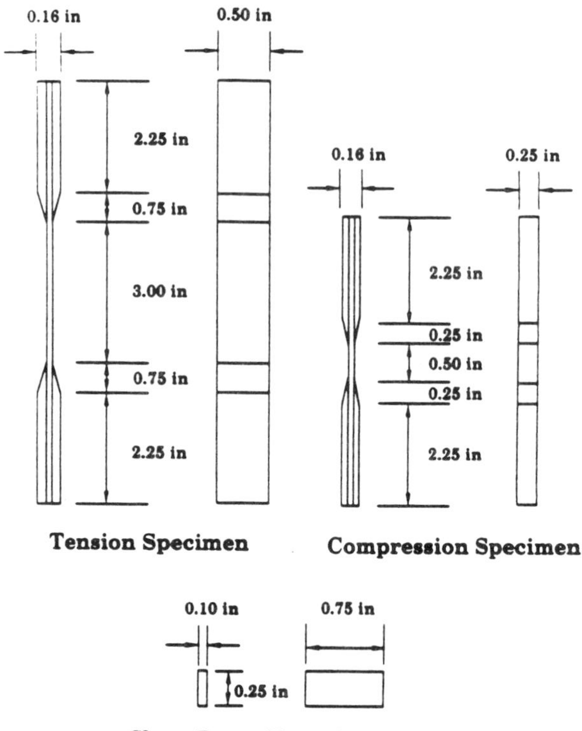

Figure 6.4 Illustration of the specimens used for mechanical properties tests.

6.2 TEMPERATURE

The temperatures, modified ionic conductivities and compactions measured in laminates cured by SECURE are given in Figures 6.5-6.10. The results in these figures show that SECURE performed all of its required tasks very well, namely

1) The autoclave, laminate midpoint and surface temperatures stayed below the maximum allowed value (Eq. 6.1).
2) With some exceptions, the temperature difference ΔT^H_{max} remained below the maximum allowed value. Only for the 200 and 1000 ply laminates did, on two occasions, ΔT^H_{max} exceed the allowed limit. At the start of the cure which was permitted (see Section 4.1) since there was no danger of an exotherm, and at the time

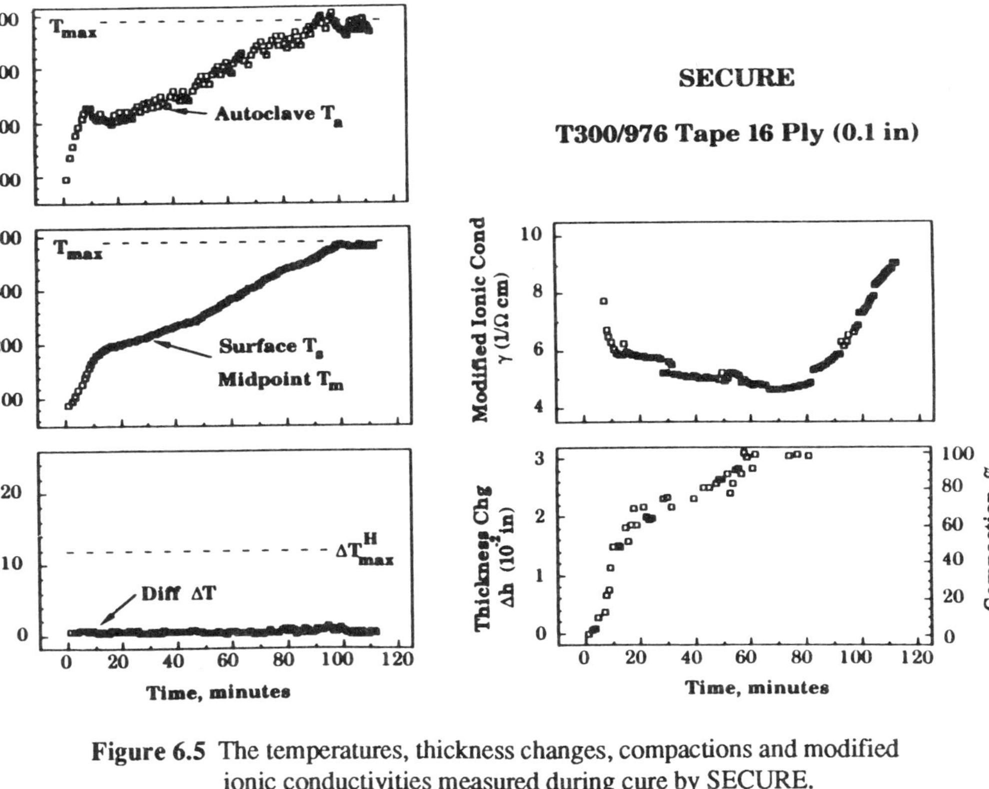

Figure 6.5 The temperatures, thickness changes, compactions and modified ionic conductivities measured during cure by SECURE.

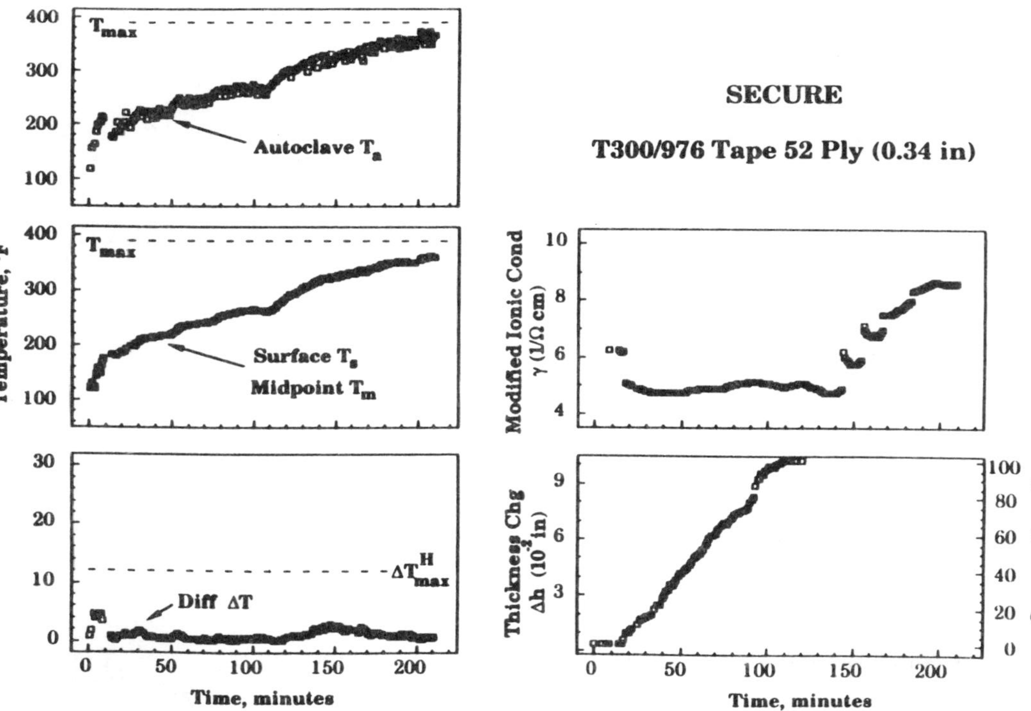

Figure 6.6 The temperatures, thickness changes, compactions and modified ionic conductivities measured during cure by SECURE.

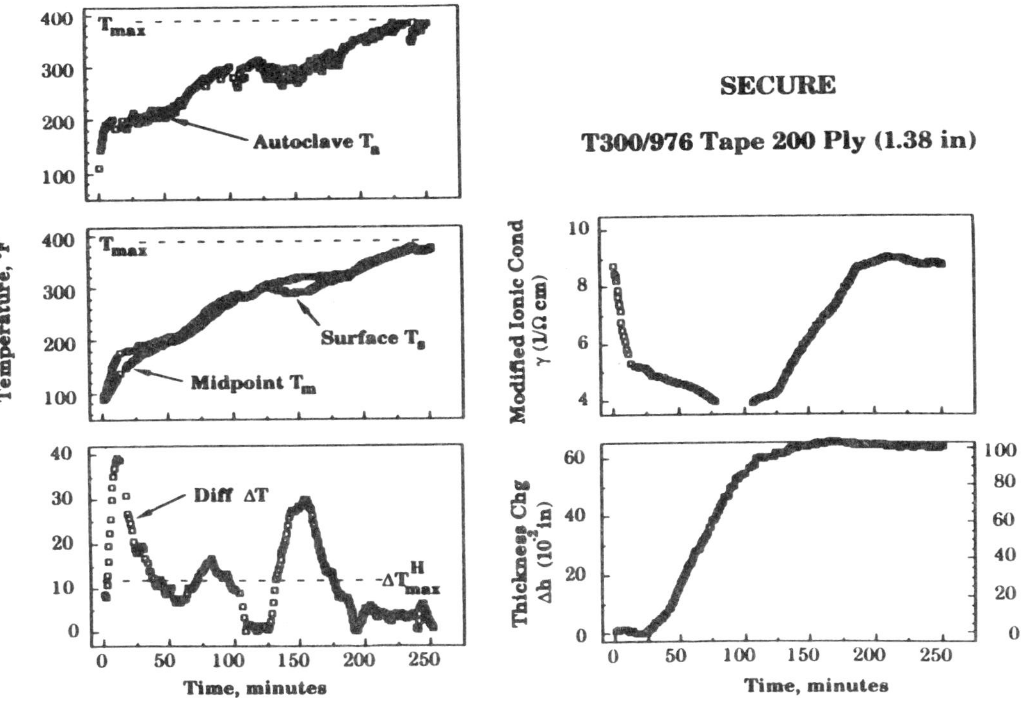

Figure 6.7 The temperatures, thickness changes, compactions and modified ionic conductivities measured during cure by SECURE.

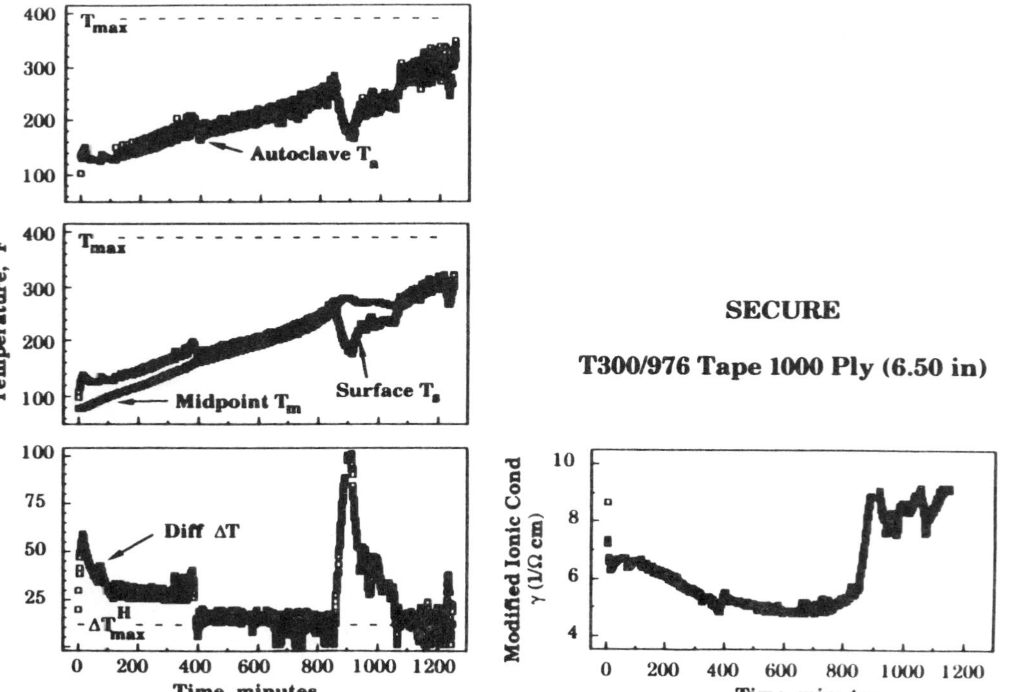

Figure 6.8 The temperatures, thickness changes, compactions and modified ionic conductivities measured during cure by SECURE.

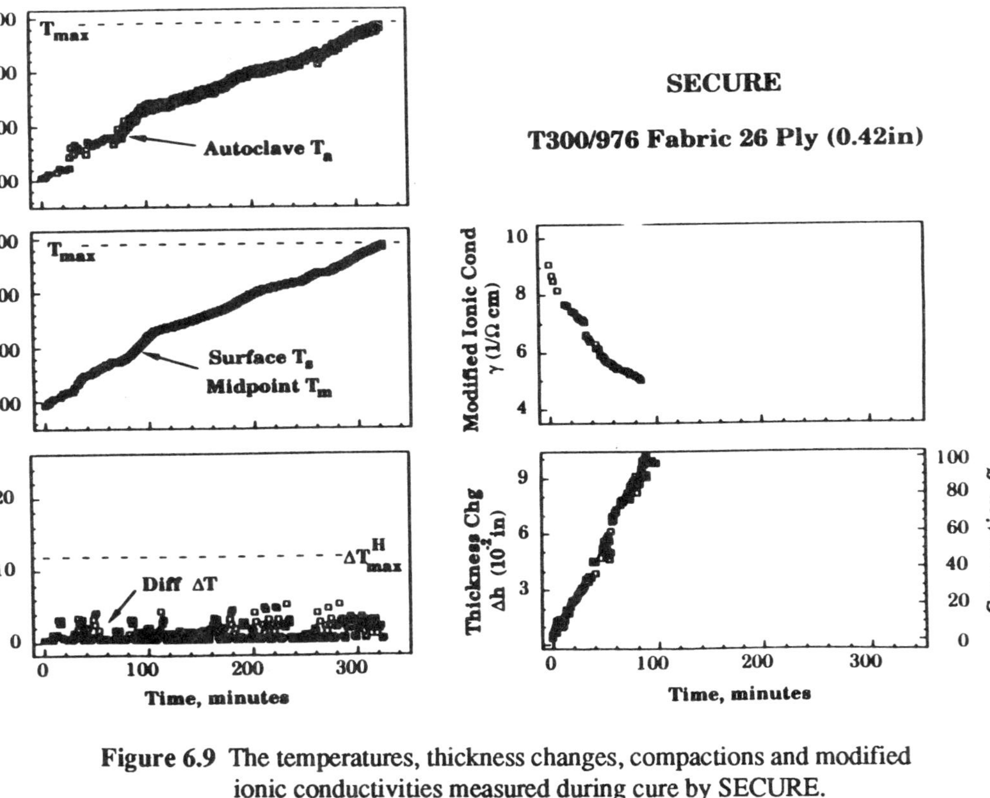

Figure 6.9 The temperatures, thickness changes, compactions and modified ionic conductivities measured during cure by SECURE.

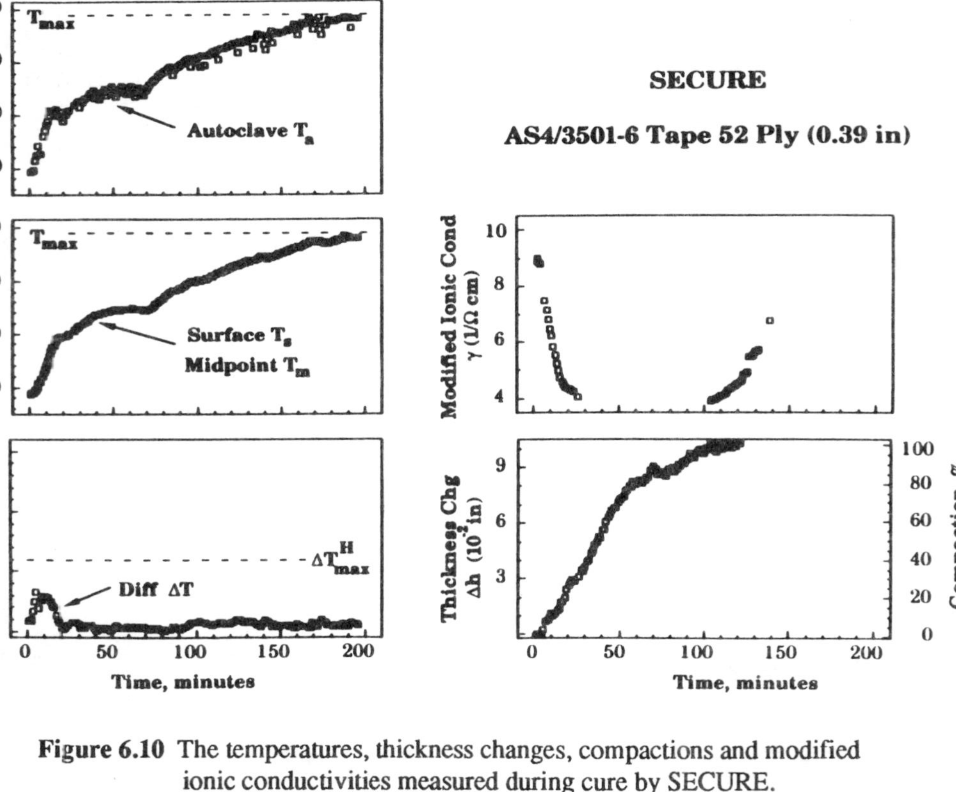

Figure 6.10 The temperatures, thickness changes, compactions and modified ionic conductivities measured during cure by SECURE.

of the exotherm. In both instances SECURE brought the temperature difference under control and to within the prescribed limits.

3) The modified ionic conductivity decreased rapidly to a low value at which it remained until the compaction process was completed. The modified ionic conductivity then increased until the laminate was fully cured i.e. until $d\gamma/dt$ became zero. For the T300/976 fabric laminate the dielectric sensor failed after approximately 100 minutes. Even in the absence of the sensor full compaction was achieved.
4) Every laminate was fully compacted.

The temperature, modified ionic conductivity and compaction were also measured in laminates cured by the optimized conventional cure cycles. The data are presented in Appendix F. It is merely pointed out here that these temperature, modified ionic conductivity and compaction data followed closely the predictions of the CURE computer code.

Although the SECURE and conventional methods result in acceptable cures, it is noteworthy that the cure cycle recommended by the material supplier ("manufacturers recommended cure cycle") may result in unacceptable temperatures. To illustrate this, the midpoint temperatures in a 200 ply T300/976 laminate were calculated by the CURE computer code for the autoclave temperature specified by the manufacturer. The results given in Figure 6.11 show that at about 180 minutes there is a significant exotherm resulting in an excessive temperature rise.

6.3 CURE TIME

The cure times (i.e. the time required to reach 100 percent cure, ($d\alpha / dt = d\gamma / dt = 0$) are listed in Table 6.1 and shown in Figure 6.12. In this figure the cure times are shown for laminates cured by SECURE as well as by the optimized conventional cure cycles. It is seen that SECURE offers savings in cure time over the optimized conventional cure cycle. Since the optimized conventional cure is one of the shortest cure cycles, SECURE would most likely yield even greater time savings over conventional cure cycles which are not "optimized".

6.4 MECHANICAL PROPERTIES

The tensile, compression and short beam shear properties of $[0_2/45_2/90_2]_s$ specimens taken from each laminate were measured. The average tensile and compression strengths and modulii and short beam shear

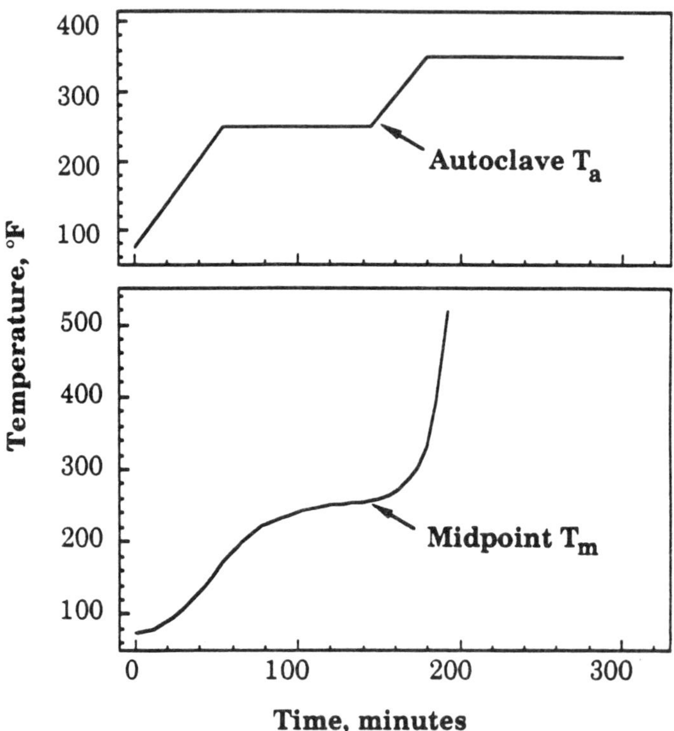

Figure 6.11 The midpoint temperatures calculated by the CURE code for the autoclave temperature profile recommended by the manufacturer.

Table 6.1

The Cure Time of Laminates Cured by SECURE and Optimized Conventional Cures

Material	No. Plies	Cure Time (min) SECURE	Conventional
Fiberite T300/976	16	112	94
(HYE-3076E)	52	211	320
6K Tape	200	252	350
	1000	1255	
Fiberite T300/976			
HMF 349D/76	26		320
6K Woven Fabric			
Hercules			
AS4/3501-6	52	196	300
(6K Tape)			

strengths are summarized in Table 6.2, and are plotted in Figures 6.13-6.15 for laminates made of tape and in Figure 6.16 for laminates made of fabric. Each data is the average of three measurements. The spread in the tensile, compressive and short beam shear data were about 5, 12, and 25 percent, respectively.

Two important observations can be made from Figures 6.13-6.15. First, the mechanical properties of laminates cured by SECURE and by the optimized conventional cure cycles are comparable. The modulii and strengths of laminates cured by SECURE are as high as those of the laminates cured by optimized conventional cures. The one exception seems to be the short beam shear strength of the 16 ply T300/976 laminate. This is likely to be due to inaccuracies in the measurements, which for short beam shear strengths were as high as 25 percent. The second important observation to be made from Figures 6.13-6.15 is that the mechanical properties do not change as the thickness of the laminate increases.

It is also noted that compression strengths of specimens made of fabric are lower than those of specimens made of tape (Figure 6.16). Nevertheless, the mechanical properties of the fabric specimens made by SECURE and by

Table 6.2

The Compressive, Tensile, and Short Beam Shear Properties of $[0_2/\pm45_2/90_2]_s$ Specimens Taken from Laminates Cured by SECURE and by Optimized Conventional Cure Cycles

(strength in ksi, modulus in msi)

		Tension				Compression				SBS	
		SECURE		Conventional		SECURE		Conventional		SECURE	Conv
	# Plys	Strength	Modulus	Strength	Modulus	Strength	Modulus	Strength	Modulus	Strength	Strength
Fiberite T300/976 6K Tape	16	51.50	7.38	53.20	7.43	62.50	7.36	50.30	6.88	4.80	7.60
	52	53.00	7.32	52.70	7.17	61.50	6.57	62.00	6.47	7.50	8.50
	200					58.70	6.83	62.10	7.13	7.80	4.80
	1000					60.30	7.16			7.10	
Fiberite T300/976 6K Fabric	26	58.50	7.09	53.00	7.70	44.80	6.11	42.80	6.66	7.50	5.70
Hercules AS4/3501-6 6K Tape	52	71.70	6.38	69.30	5.92	68.60	6.41	73.50	6.63	5.30	5.90

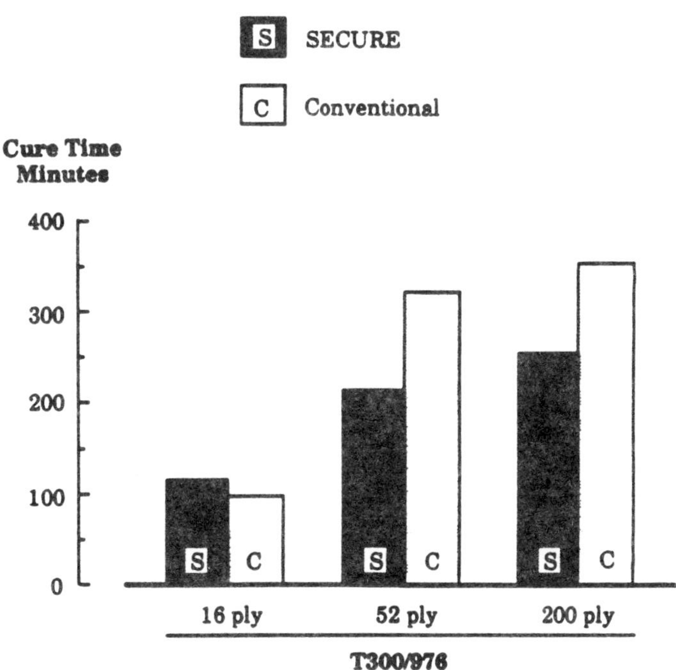

Figure 6.12 The cure times of laminates cured by SECURE and by optimized conventional cure cycles.

the optimized conventional cure cycles are nearly the same.

The void contents and fiber volumes (the average of two samples) of laminates cured by SECURE and by the optimized conventional cure cycles are listed in Table 6.3 and are shown in Figures 6.17-6.19. The void contents of the laminates cured by SECURE and by optimized conventional cure cycles are very close. Furthermore, the void contents of all the laminates are very low; they are practically zero for the 52 and 200 ply T300/976 laminates, and are under 0.75 percent for all the other laminates, regardless whether they were cured by SECURE or by the optimized conventional cure cycles.

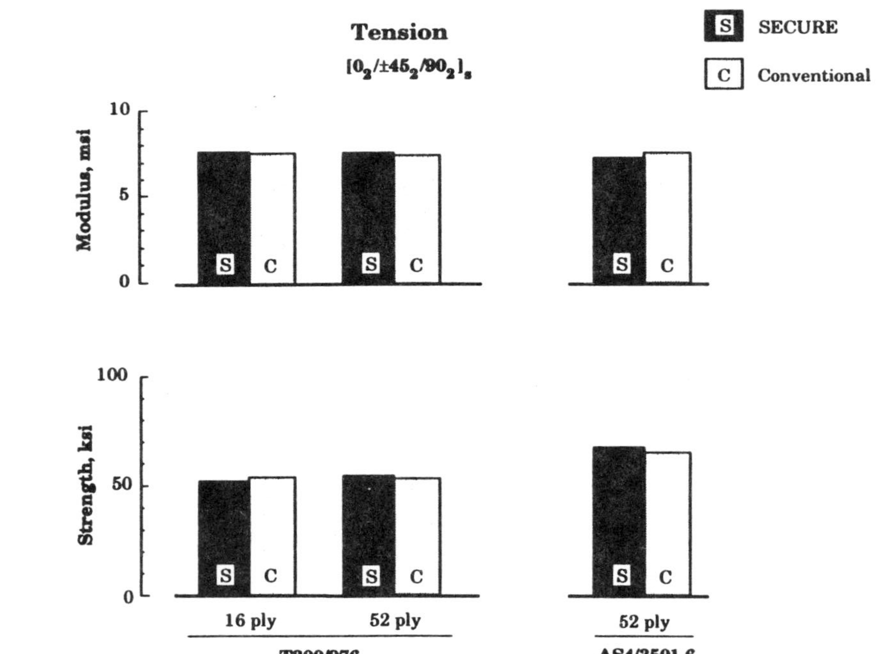

Figure 6.13 The tensile strengths and modulii of $[0_2/\pm45_2/90_2]_s$ specimens taken from 16 and 52 ply thick laminates cured by SECURE and by optimized conventional cure cycles.

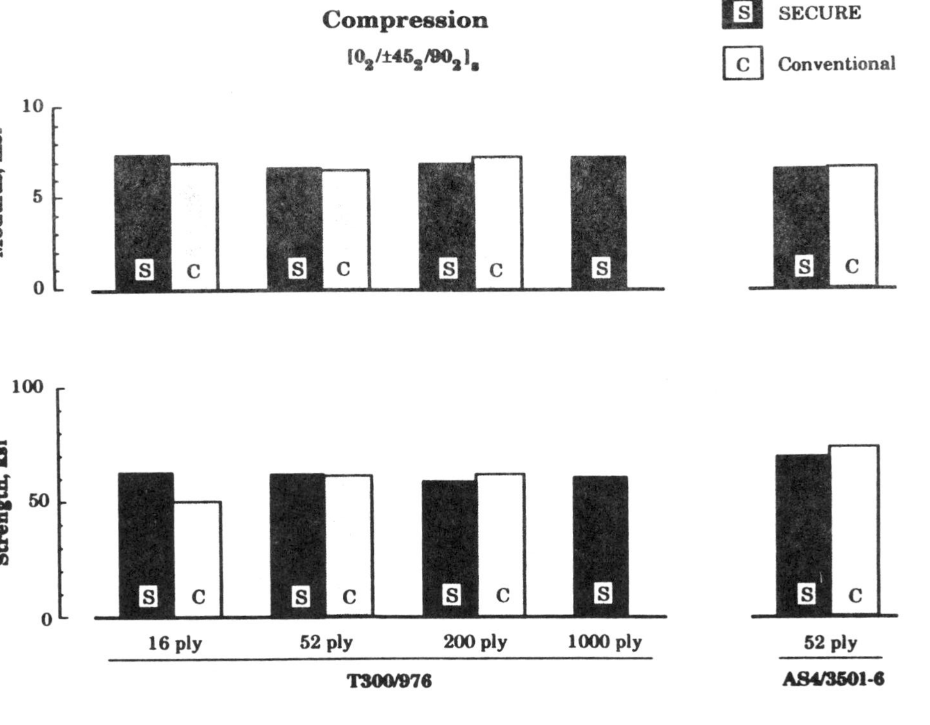

Figure 6.14 The compression strengths and moduli of $[0_2/\pm45_2/90_2]_s$ specimens taken from laminates cured by SECURE and by optimized conventional cure cycles.

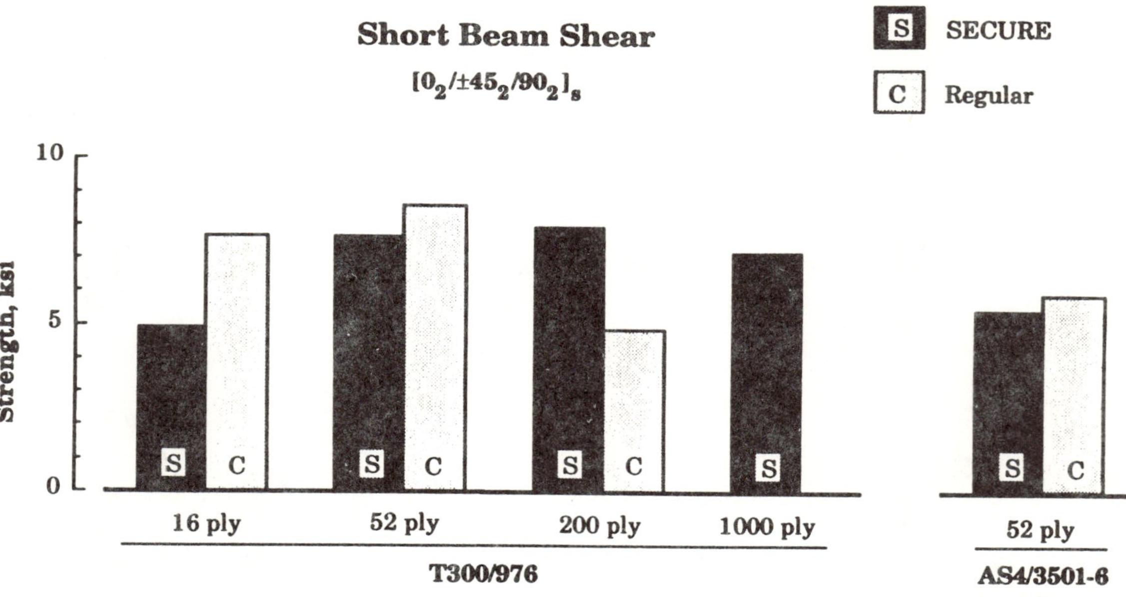

Figure 6.15 The short beam shear strengths and modulii of $[0_2/\pm45_2/90_2]_s$ specimens taken from laminates cured by SECURE and by optimized conventional cure cycles.

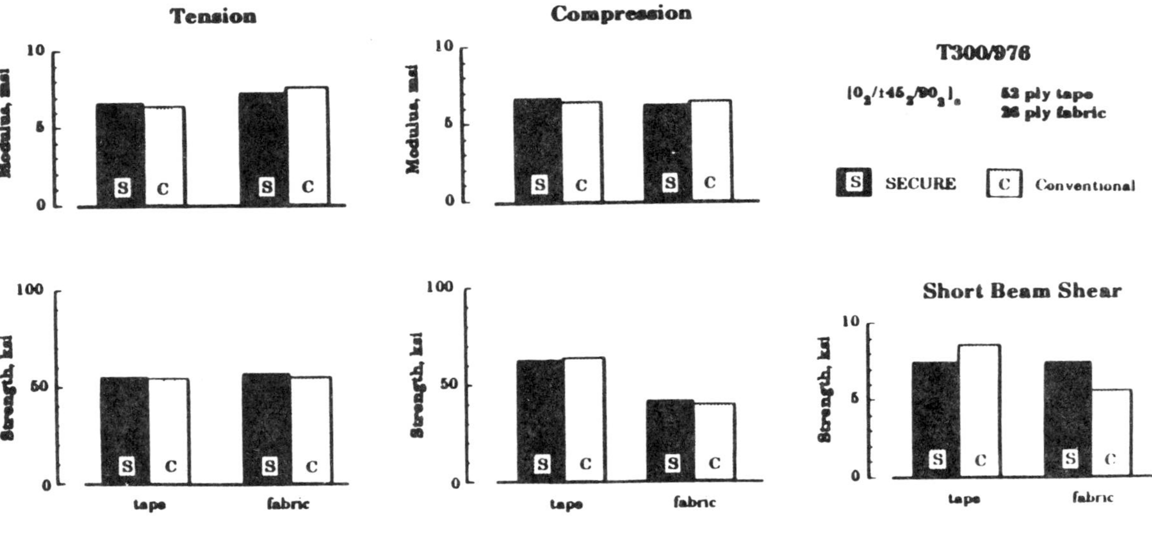

Figure 5.16 The tensile, compressive and short beam shear properties of $[0_2/\pm45_2/90_2]_s$ tape and fabric specimens taken from 26 and 52 ply laminates cured by SECURE and by optimized conventional cure cycles.

Table 6.3

The Void Contents and Fiber Volumes of Laminates Cured by SECURE and by Optimized Conventional Cures

		Void Content, %		Fiber Volume, %	
	# Plys	SECURE	Conventional	SECURE	Conventional
Fiberite T300/976 6K Tape	16	0.6	0.0	64.4	68.1
	52	0.0	0.0	60.6	63.2
	200	0.0	0.0	65.9	64.7
	1000	0.1		63.4	
Fiberite T300/976 6K Fabric	26	0.5	0.7	59.1	64.3
Hercules AS4/3501-6 6K Tape	52	0.2	0.4	59.6	59.1

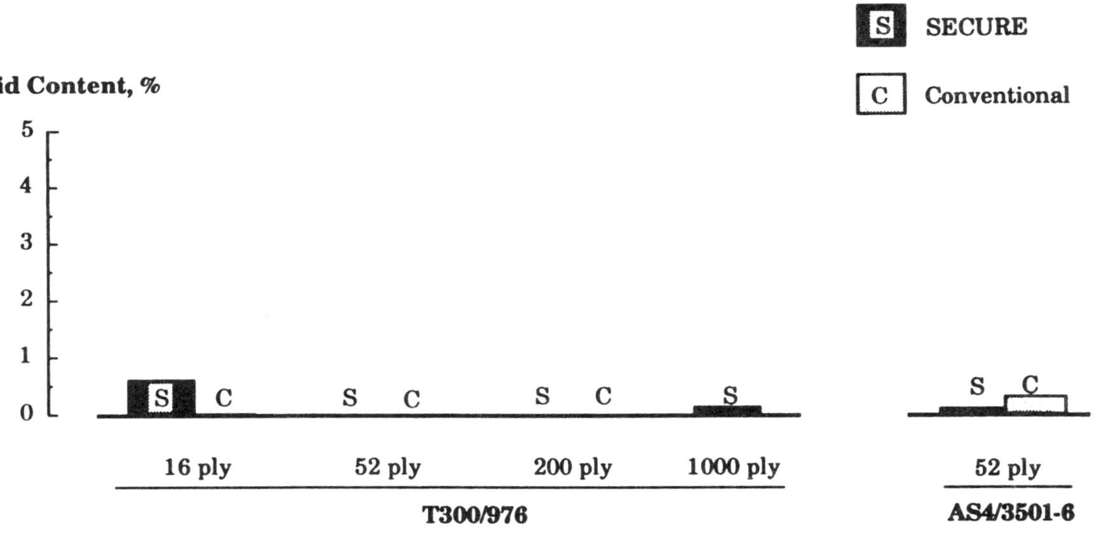

Figure 6.17 The void contents of 16, 52, 200 and 1000 ply thick laminates cured by SECURE and by optimized conventional cure cycles.

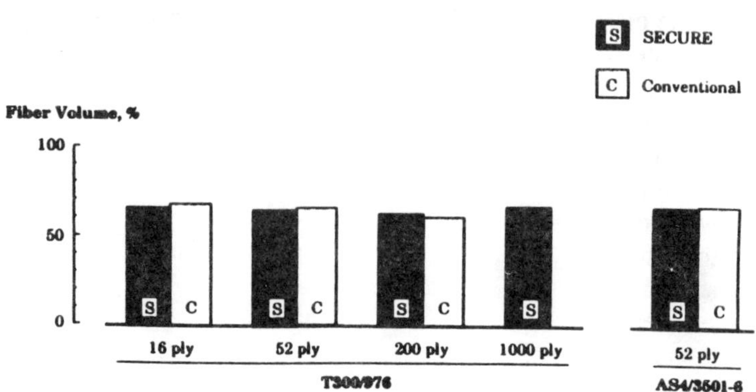

Figure 6.18 The fiber volumes of 16, 52, 200 and 1000 ply thick laminates cured by SECURE and by optimized conventional cure cycles.

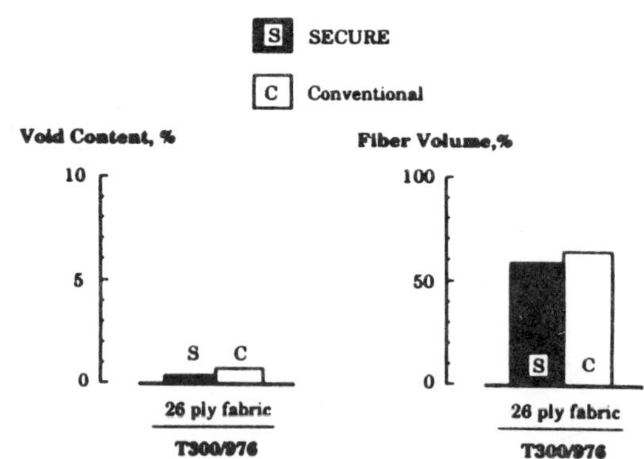

Figure 6.19 The void contents and fiber volumes of fabric specimens taken from 26 ply laminates cured by SECURE and by optimized conventional cure cycles.

Observe that the void content of the 16 ply T300/976 tape laminate cured by the optimized conventional cure cycle was practically zero, while the void content of the same type of laminate cured by SECURE was 0.6 percent. The reason for this is as follows. The autoclave pressure for the optimized conventional cure cycle was 200 psia and for the SECURE cure it was 150 psia. The saturation temperature of water vapor T_s corresponding to these pressures are 381.79°F and 358.42°F, respectively. In the laminate cured by the optimized conventional cure cycle the temperature was always below the saturation temperature of 381.79°F (see Figure 6.20, left). Thus, the saturation pressure was always less than the autoclave pressure. This pressure difference seems to have been adequate to supress the voids. In the laminate cured by SECURE the saturation temperature of 358.42°F was reached at t = 92 minutes, at which time the viscosity (as represented by the modified ionic conductivity, Figure 6.20, right) was still low. From this time on the saturation pressure was higher than the autoclave pressure. Hence, from t = 92 minutes until the resin hardened at about 105 minutes voids could form and grow. This resulted in the 0.6 percent void content.

The aforementioned experimental results are consistent with the rules formulated in Table 4.3.

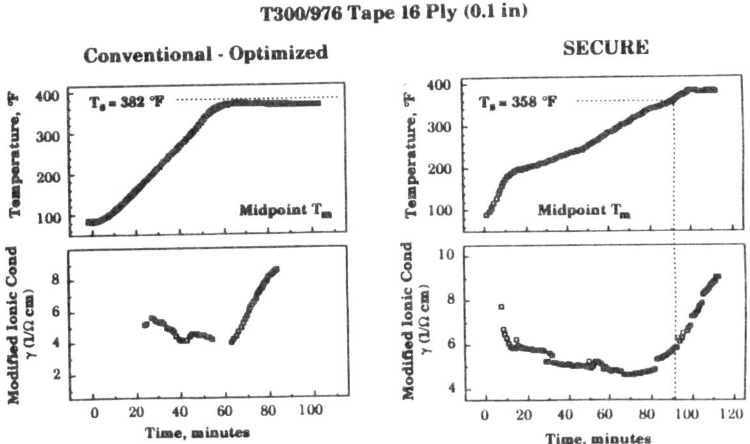

Figure 6.20 Midpoint temperature and modified ionic conductivity data for 16 ply T300/976 tape laminates cured by optimized conventional cure cycle (left) and cured by SECURE (right).

The fiber volume fractions are nearly the same for laminates cured by SECURE and by optimized conventional cure cycles and are in the range of 59 to 68 percent. This is within the range (55 - 68 percent) usually specified for these materials.

The aforementioned data show that SECURE performs its functions well and results in high quality laminates.

6.5 RESIDUAL STRESS

Residual (curing) stresses are introduced into the laminate during cool-down. To evaluate the effect of the cure temperature on damage introduced by residual stresses, Fiberite T300/976 crossply $[0_4/90_4]_S$ laminates were cured under the temperature cycles shown in Figure 6.21 at a pressure of 100 psia. Some of the laminates were cured at a maximum temperature of $T_{max} = T_{crit} =$ 300°F, while others at $T_{max} = T_{crit} = 375$°F. The cool-down after cure was either at a "fast" (~10°F/min) or at a "slow" (~1°F/min) rate. After a laminate reached end of the temperature cycle shown in Figure 6.21, it was taken out

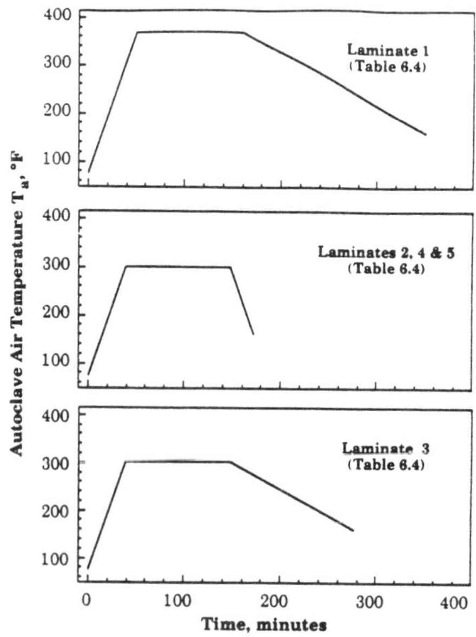

Figure 6.21 The temperature cycles used to evaluate the effect of the cure temperature on damage.

Table 6.4

Matrix Cracking in Fiberite T300/976 Cross-Ply $[0_4/90_4]_s$ Laminates
(Cure Cycles Shown in Figure 6.21)

Laminate	Max Cure Temp., (°F)	Cooling Rate (°F/min)	Degree of Cure After Cure	Damage (Cracks)	Post Cure Time at 375 °F, (min)	Degree of Cure After Post Cure	Damage (Cracks)
1	375	1	0.94	yes			
2	300	10	0.71	no	90	0.94	yes
3	300	1	0.73	no	90	0.94	yes
4	300	10	0.72	no	3	0.76	yes
5	300	10	0.72	no	90	0.94	yes

of the autoclave, air cooled to room temperature, X-rayed, and inspected for damage.

After X-ray, the laminates cured at 300°F were placed in an oven at 375°F and postcured for different lengths of time. One was kept in the oven for 3 minutes. During this time its temperature increased to 375°F while (according to results computed by the CURE code) its degree of cure increased only from $\alpha = 0.72$ to $\alpha = 0.76$. Two of the laminates were kept in the oven for 90 minutes, during which time they became completely cured ($\alpha \to 1$). After the oven cure each laminate was air cooled to room temperature and was again X-rayed.

The results of these tests are summarized in Table 6.4 The results show that laminates heated to only 300°F did not have matrix cracks, even when cooled "fast" after the cure. On the other hand, laminates whose temperature was raised to 375°F had significant matrix cracking, regardless whether they were cooled "fast" or "slow". Even when the degree of cure was only 76 percent ($\alpha = 0.76$) matrix cracking occurred when the cross-ply laminate's temperature was raised to 375°F. In contrast, the unidirectional and quasiisotropic laminates (Table 4.1) did not show signs of matrix cracking at such high cure temperatures.

The major conlcusion suggested by these results is that matrix cracking does not occur if the maximum temperature is kept under a critical value ($T_{max} \leq T_{crit}$). The value of T_{crit} depends strongly on the layup and the material.

PART 2

Processing Thermoplastics

CHAPTER 7

General Considerations for Processing Thermoplastics

Processing of thermoplastic matrix composites consists of three major steps. First, matrix is introduced into the space between the fibers in each tow, "impregnating" the tow. Second, individual plies are "consolidated" such that there is good contact as well as good bond between the fibers. Third, the composite is heated and then cooled at a rate which provides the desired "crystallinity" in the matrix. The conditions which must be achieved during processing are summarized in Table 7.1.

The aforementioned processes have been described by a model, which consists of three submodels. These submodels correspond to the three processing steps of impregnation, consolidation, and crystallinity. A model encompassing all three submodels has been presented by Lee and Springer [12]. Submodels for impregnation and consolidation have been proposed respectively by Kim et al [33] and by Dara and Loos [13]. Submodels representing the crystallization process have been developed by Lee et al [12, 14] and by Seferis and Velisaris [15, 16].

Table 7.1

Conditions Which Must Be Met During Processing of Thermoplastic Matrix Composites
• The tows must be fully impregnated.
• The individual plies must be in complete intimate contact.
• The bonds between individual plies must be complete.
• The crystallinity must have the desired value throughout the composite.

CHAPTER 8

A Model for Processing Thermoplastics

8.1 IMPREGNATION

Impregnation of the fiber tows with resin is a difficult process because of the high matrix viscosity. Generally, impregnation is best accomplished by prepreg manufacturers. Therefore, impregnation models are not given here in detail. It is merely noted that Lee and Springer [12] proposed a model of impregnation based on the concept that the matrix penetrates into the fiber tow under the action of fiber tension. Kim et al [33] presented a model which describes the penetration of the matrix due to applied pressure.

8.2 CONSOLIDATION AND BONDING

When thermoplastic composites are being processed, individual plies consolidate into a laminate by bonding at the interfaces [12, 13]. This bonding consists of two phenomena. First, two adjacent ply surfaces coalesce and come into "intimate" contact". Second, bond forms at the ply interface by a process called autohesion. The following models describe the effects of the processing variables (temperature, pressure, time) on the intimate contact and autohesion processes. These processes occur simultaneously. However, for the sake of clarity, the models of these processes are developed separately.

Intimate Contact

Laminates are formed by "laying up" plies. Since the ply surfaces are uneven, spatial gaps exist between the plies prior to the application of heat and pressure. When heated, most thermoset matrix composites, including epoxy, have sufficiently low viscosities and wetting abilities to coalesce the ply surfaces. For thermoplastics, even at elevated temperatures, the viscosity is too high to produce the desired degree of flow along the interface. Therefore, thermoplastic matrix composites must actually be deformed to produce intimate contact between adjacent surfaces.

The irregular ply surface is represented by a series of rectangles (Figure 8.1). Dara and Loos utilized rectangles of different sizes. Here the rectangles are taken to be identical. Due to the applied force (or pressure) the rectangular elements spread along the interface. We are interested in the extent of this spread (degree of intimate contact) as a function of applied temperature, force (pressure), and time. We also wish to calculate the time required to reach "complete" intimate contact.

With reference to Figure 8.2, the degree of intimate contact is defined as

$$D_{ic} = \frac{b}{w_o + b_o} \tag{8.1}$$

where b_o and b are the initial ($t \leq 0$) and instantaneous (at time t) widths of each rectangular element, respectively, and w_o is the initial distance between two adjacent elements. During processing the volume of each element remains constant

$$V_o = a_o b_o = ab \tag{8.2}$$

where a_o and a are the initial and instantaneous heights of each rectangular element. Equations 8.1 and 8.2 yield the following expression for the degree of intimate contact

$$D_{ic} = \frac{a_o / a}{1 + w_o / b_o} \tag{8.3}$$

To proceed with the model we apply the law of conservation of mass to a control volume of width dy. Referring to Figure 8.2 we write

$$a\frac{du_y}{dy} + \frac{da}{dt} = 0 \tag{8.4}$$

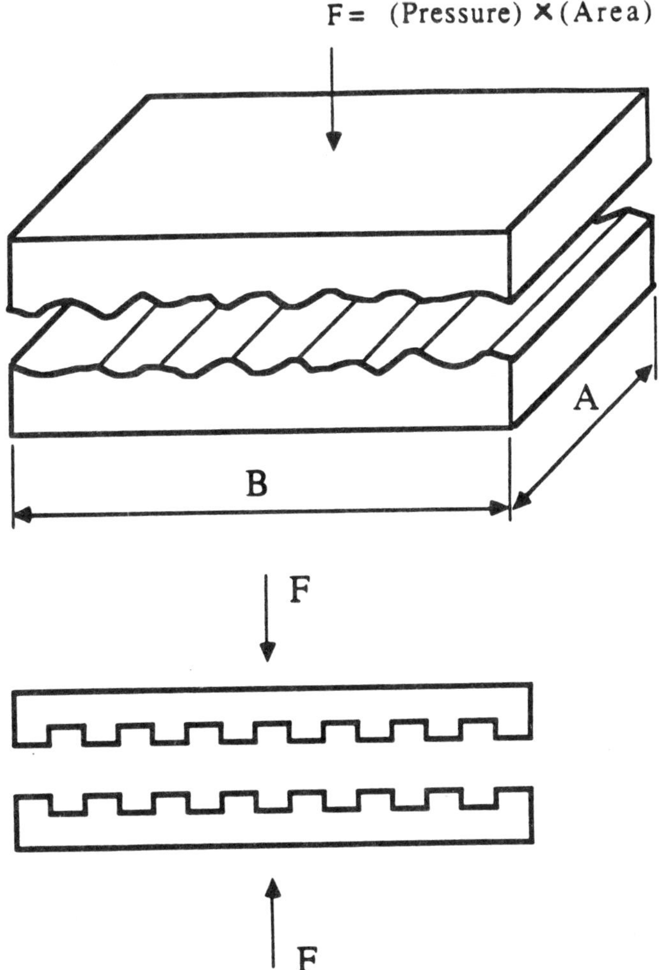

Figure 8.1 Illustration of the idealized interface used in the intimate contact model.

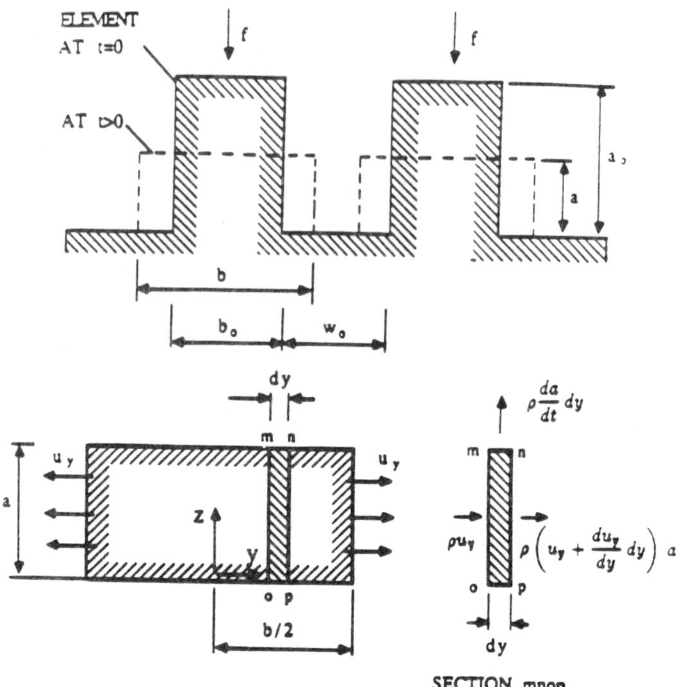

Figure 8.2 Top: Rectangular elements representing the uneven surface at time $t = 0$. Bottom: Illustration of one element at time t, and the control volume used in calculating mass flows.

y is the coordinate along the interface and t is time. By assuming that the flow is laminar, we write the average velocity u_y as [26]

$$u_y = -\frac{a^2}{12\mu_{mf}} \frac{dP}{dy} \tag{8.5}$$

Here μ_{mf} is the viscosity of the fiber-matrix mixture. In the space between two adjacent elements the pressure is P_e which is taken to be equal to the ambient pressure. The edge of the element ($y = b/2$), moves with a speed of db/dt. Thus, the boundary conditions corresponding to Eq. 8.5 are

$$P = P_e \text{ and } u_y = \frac{db}{dt} \quad \text{at } y = b/2 \text{ for } t > 0 \tag{8.6}$$

By combining Eqs. 8.4-8.6, after algebraic manipulations, we obtain

$$P - P_e = \frac{6\mu_{mf}}{a^3}\frac{da}{dt}\left(y^2 - \left(\frac{b}{2}\right)^2\right) \tag{8.7}$$

The force applied to the entire ply of length A and width B is F (Figure 8.1). Correspondingly, the force applied pre unit length to one element is

$$f = \frac{F}{A}\frac{1}{n} \tag{8.8}$$

where $n = B/(b_o + w_o)$ is the number of elements in a ply of width B. We write Eq. 8.8 as

$$f = \frac{F}{A}\frac{(b_o + w_o)}{B}(P_{app})(b_o + w_o) \tag{8.9}$$

where P_{app} is the applied gauge pressure. The force applied to an element must be balanced by the pressure inside the element

$$f = \int_{-b/2}^{b/2}(P - P_e)dy \tag{8.10}$$

Equations 8.2, 8.7, 8.9, and 8.10 yield

$$\frac{a_o}{a} = \left[1 + \frac{5P_{app}}{\mu_{mf}}\left(1 + \frac{w_o}{b_o}\right)\left(\frac{a_o}{b_o}\right)^2 t\right]^{1/5} \tag{8.11}$$

Equations 8.11 and 8.3 provide the following expression for the degree of intimate contact

$$D_{ic} = \frac{1}{1+\dfrac{w_o}{b_o}}\left[1+\frac{5P_{app}}{\mu_{mf}}\left(1+\frac{w_o}{b_o}\right)\left(\frac{a_o}{b_o}\right)^2 t\right]^{1/5} \tag{8.12}$$

Complete intimate contact is achieved when D_{ic} becomes unity. The height a corresponding to this condition is (see Eq. 8.3)

$$a_{D_{ic=1}} = \frac{a_o}{1+w_o/b_o} \tag{8.13}$$

Substitution of Eq. 8.13 into Eq. 8.12 results in the following expression for the time required to achieve complete intimate contact

$$t_{ic} = \frac{\mu_{mf}}{5P_{app}}\frac{1}{1+w_o/b_o}\left(\frac{b_o}{a_o}\right)^2\left[\left(1+\frac{w_o}{b_o}\right)^5 - 1\right] \tag{8.14}$$

The degree of intimate contact and the time required to reach complete intimate contact can be calculated by Eqs. 8.12 and 8.14 respectively. The geometric parameters w_o/b_o and a_o/b_o can be measured from photomicrographs of the cross section of an uncompacted ply. The viscosity μ_{mf} can be obtained by matching the degree of intimate contact model to data.

Autohesion

Once two adjacent interfaces come into intimate contact, a bonding process between the interfaces starts. In the case of two similar thermoplastic interfaces the bond is formed mostly by autohesion. During autohesion segments of the chain like molecules diffuse across the interface (Figure 8.3). The extent of the molecular diffusion and, hence, the bond strength increase with time.

A convenient way of characterizing the extent of autohesion is through the bond strength. Thus a degree of autohesion may be defined as [13]

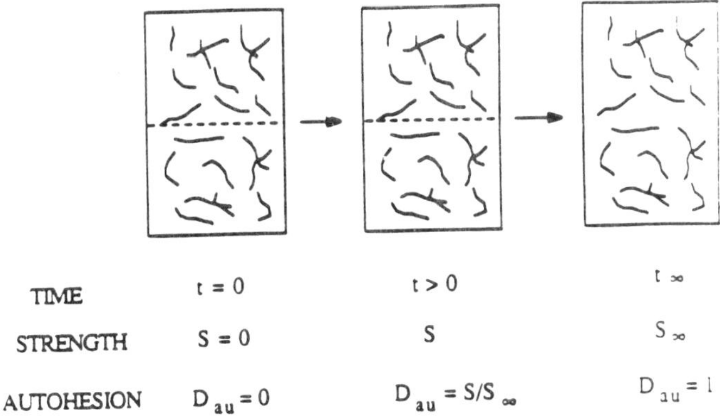

Figure 8.3 Illustration of the autohesion process.

$$D_{au} = \frac{S}{S_{\infty}} \tag{8.15}$$

where S is the bond strength at time t and S_{∞} is the ultimate bond strength, i.e., the strength of a completely bonded interface.

We approximate the autohesion by the following expression

$$D_{au} = \kappa t_a^{1/4} \tag{8.16}$$

where t_a is the time elapsed from the start of the autohesion process (i.e., the time when the interfaces come into intimate contact). κ is a constant which is related to the temperature through the Arrhenius relation

$$\kappa = \kappa_o \exp(-E / RT) \tag{8.17}$$

κ_o is a constant, E is the activation energy and R is the universal gas constant. It is noted that the crystallinity of the material resulting from cooling may affect the accuracy of Eq. 8.16.

The time required to complete the autohesion process is discussed in the next section.

Degree of Bond

Once intimate contact is established at a point along the interface autohesion, and thus the bonding process, starts. The degree of bonding D_b at time t can be calculated by the expression

$$
\begin{aligned}
(D_b)_{at\,t} = &\left[(D_{ic})_{at\,\Delta t} - (D_{ic})_{at\,0}\right]\left[\kappa(t-\Delta t)^{1/4}\right] \\
&+\left[(D_{ic})_{at\,2\Delta t} - (D_{ic})_{at\,\Delta t}\right]\left[\kappa(t-2\Delta t)^{1/4}\right]+\ldots
\end{aligned}
\tag{8.18}
$$

$$
(D_b) = \sum_{i=1}^{t/\Delta t}\left[(D_{ic})_{i\Delta t} - (D_{ic})_{(i-1)\Delta t}\right]\left[\kappa(t - i\Delta t)^{1/4}\right] \tag{8.19}
$$

where Δt is an arbitrary time step. D_{ic} is the degree of intimate contact given by Eq. 8.12.

Bonding is complete when D_b becomes unity. Thus the time required to complete the bonding t_b is calculated by

$$
1 = \sum_{i=1}^{t_b/\Delta t}\left[(D_{ic})_{i\Delta t} - (D_{ic})_{(i-1)\Delta t}\right]\left[\kappa(t_b - i\Delta t)^{1/4}\right] \tag{8.20}
$$

The parameters required for the consolidation submodel are listed in Table 8.1

8.3 CRYSTALLINITY

For semicrystalline thermoplastic matrix composites the degree of crystallinity affects significantly the mechanical properties of the composite. The desired crystallinity is achieved during the processing by cooling the composite from the melt temperature T_{melt} at an appropriate cooling rate. The objective here is to establish a model which relates the cooling rate applied during processing to the crystallinity of the material. For clarity, the model is developed for a flat plate in which the temperature varies only across the plate but not along the plate (one dimensional problem, Figure 8.4).

Table 8.1

Input Parameters Required for the Consolidation Submodel
Intimate Contact
Length of the plate
Width of the plate
Geometric ratios a_o/b_o and b_o/w_o (see Figure 8.2)
Matrix fiber viscosity
Force (or gauge pressure) applied to the plate
Applied temperature
Autohesion
Porportionality constant κ (see Eq. 8.17)
Applied temperature

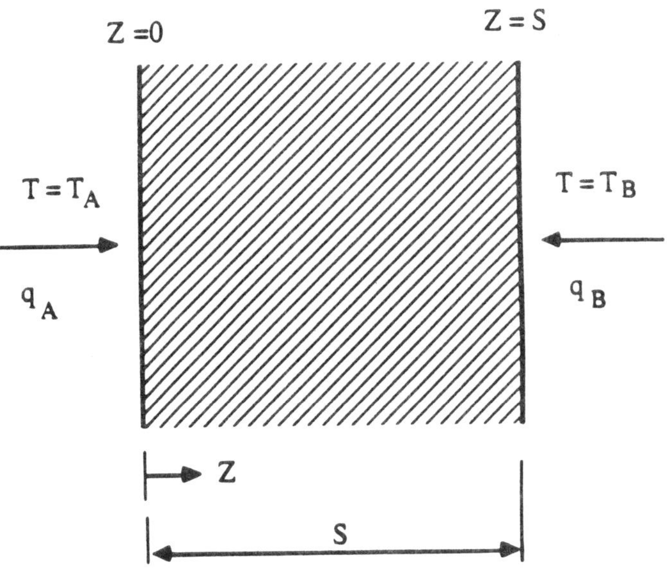

Figure 8.4 Geometry used in the crystallinity submodel.

During cooling the crystallinity of the polymer changes. The instantaneous degree of crystallinity depends on the temperature and on the rate of change of the temperature. Therefore, to determine the degree of crystallinity as a function of position and time, the temperature distribution inside the laminate must be known at all times. For a flat plate the temperature distribution can be calculated by the following form of the conservation of energy [34]

$$\rho C \frac{\partial}{\partial t} = \frac{\partial}{\partial z}\left(k \frac{\partial T}{\partial z}\right) + m_m \frac{dc}{dt} H_u \tag{8.21}$$

where t is time, z is the coordinate normal to the plate, T is the temperature, ρ is the density, C is the specific heat, and k is the thermal conductivity of the composite. Expressions for estimating the latter three parameters are given in Table 8.2.

The second term on the right-hand side of Eq. 8.21 represents the heat generated due to crystallization. In this term c is the crystallinity of the matrix, m_m is the matrix mass fraction, and H_u is the theoretical ultimate heat of crystallization of the polymer at 100 percent crystallinity.

Equation 8.21 contains two dependent variables, the temperature T and the crystallinity c. Therefore, to proceed with the solution an additional expression is needed relating the crystallinity to the temperature. This expression is established by utilizing the fact that for a given material, the rate of degree of crystallinity depends on the cooling rate and on the instantaneous temperature. This dependence can be expressed symbolically as

$$\frac{dc}{dt} = g\left(\frac{dT}{dt}, T\right) \tag{8.22}$$

The function g is a material property which must be obtained experimentally. A procedure for determining this function is described in Appendix G.

Solutions to Eqs. 8.21 and 8.22 require that the initial and boundary conditions be specified. Initially, (prior to the start of cooling) the composite is at the uniform temperature T_i having zero crystallinity. Thus, the initial conditions corresponding to Eq. 8.21 are

Table 8.2

Input Parameters Required for the Crystallinity Submodel
Geometry
Number of plies in the plate the plate
Thickness of one ply
Polymer[a]
Mass fraction
Density
Specific heat
Thermal conductivity
Theroretical ultimate heat of crystallization
Relationship between crystallinity, cooling rate, and temperature
Fiber[a]
Density
Specific heat
Thermal conductivity
Processing Variables
Applied temperature or heat flux
Mechanical Properties
Mechanical properties as a function of crystallinity (optional)

[a]The density, specific heat and thermal conductivity of the composite may be approximated by the expressions

$$\rho = \upsilon_m \rho_m + (1-\upsilon_m)\rho_f$$
$$C = m_m C_m + (1-m_m)C_f$$
$$k = \left[\left(1-\sqrt{1-\upsilon_m}\right) + 1/\left(\sqrt{1/(1-\upsilon_m)} + \left(k_m/k_f - 1\right)\right)\right]k_m$$

where the subscripts m and f refer to the matrix and fiber respectively. υ_m is the matrix volume fraction

$$\upsilon_m = \left[1 + \left(\rho_m/\rho_f\right)\left(1/m_m - 1\right)\right]^{-1}$$

Initial Conditions:

$$\left.\begin{array}{l} T = T_i \\ c = 0 \end{array}\right\} \begin{array}{l} t < 0 \\ 0 \le z \le s \end{array} \tag{8.23}$$

where s is the thickness of the plate. During processing either the surface temperatures (T_A and T_B) or the heat fluxes (q_A and q_B) on the two surfaces of the plate must be specified (Figure 8.4). In terms of the surface temperatures the boundary conditions are

Boundary Conditions (Surface Temperatures):

$$\begin{array}{lll} T = T_A & z = 0 & t \ge 0 \\ T = T_B & z = s & t \ge 0 \end{array} \tag{8.24}$$

In terms of the heat fluxes the boundary conditions are

Boundary Conditions (Heat Flux):

$$\begin{array}{lll} q = q_A & z = 0 & t \ge 0 \\ q = q_B & z = s & t \ge 0 \end{array} \tag{8.25}$$

The temperature and the crystallinity as functions of position and time can be calculated by Eqs. 8.21 to 8.25. The solutions to these equations require numerical procedures. The input parameters required for the crystallinity submodel are given in Table 8.2

Mechanical Properties

The mechanical properties of thermoplastic matrix composites can be related directly to the crystallinity of the polymer. The variations of the mechanical properties with the crystallinity are to be established by tests. The mechanical properties resulting from a given cooling rate can then simply be determined by calculating the crystallinity by the model and the mechanical property from the known crystallinity-property relationship. Mechanical properties of PEEK 150P polymer and APC-2 composite as functions of crystallinity have been reported in reference [14] and, for convenience, are given in Table 8.3

Table 8.3

Mechanical Properties of PEEK 150P and Unidirectional APC-2 Composite as Functions of Crystallinity

PEEK 150P

Tensile strength:	$S_t = 8.2 + 0.17\ c$ (ksi)
Tensile modulus:	$E_t = 404 + 4.0\ c + 0.075\ c^2$ (ksi)
Shear strength:	$S_s = 4.9 + 0.10\ c + 0.0015\ c^2$ (ksi)
Shear modulus:	$E_s = 140 + 1.59\ c$ (ksi)
Compression strength:	$E_t = 18 + 0.22\ c$ (ksi)
Fracture toughness:	$K = 27.6 \times 10^{-0.027c}$ (ksi)$\sqrt{\text{in}}$

APC-2

Mode I and mode II fracture toughnesses (K_{IC} and K_{IIC}) and fracture energies (G_{IC} and G_{IIC})

$K_{IC} = 4.3 - 0.007\ c - 0.001\ c^2$ (ksi)$\sqrt{\text{in}}$
$K_{IIC} = 9.2 - 0.005\ c - 0.002\ c^2$ (ksi)$\sqrt{\text{in}}$
$G_{IC} = 12.01 - 0.022\ c - 0.006\ c^2$ (lbf/in)
$G_{IIC} = 12.78 - 0.024\ c - 0.004\ c^2$ (lbf/in)

8.4 METHOD OF SOLUTION OF THE CONSOLIDATION AND CRYSTALLINITY SUBMODELS

The consolidation and crystallinity submodels may be solved separately or in combination. Solution of the consolidation submodel is relatively simple, while solution of the crystallinity submodel requires more complex numerical analysis. To facilitate the solutions to these submodels, an algorithm, supported by a user-friendly computer code, was developed. This code, designated as PLASTIC, provides numerical results to both submodels. The results which can be generated by the code are summarized in Figure 8.5 and Table 8.4.

The PLASTIC code can be used to establish the appropriate variables, namely the heating and cooling rates and pressures which need to be applied

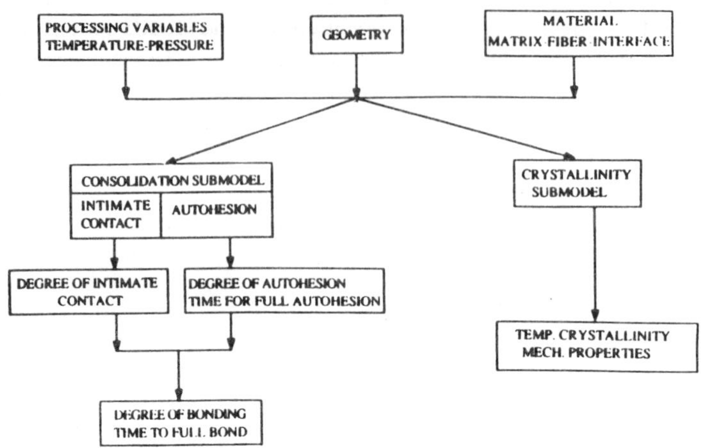

Figure 8.5 Illustration of the output of the thermoplastic processing model.

Table 8.4

Output Provided by the PLASTIC Computer Model
Consolidation Submodel
Degree of intimate contact as a function of time,
Time required to achieve complete intimate contact,
Degree of autohesion as a function of time,
Time required to achieve full autohesion,
Degree of bonding as a function of time,
Time required to achieve full bonding.
Crystallinity Submodel
Crystallinity of the matrix as a function of position and time,
Crystallinity of the matrix as a function of position at the end of the manufacturing process,
Mechanical properties of the material at the end of the manufacturing process.

during the manufacture of a part of the required quality. The procedure to be used with this code is similar to the one used with the CURE code (Chapter 3). A set of process variables are entered into the code. The results generated for these variables are examined. The calculations are repeated until the process variables are such that the conditions given in Table 7.1 are met.

APPENDICES

Methods for Cure Kinetics and Viscosity

The use of models requires that the chemical properties of the material be known. Unfortunately, a complete chemical characterization of the material during cure is a formidable task. Nevertheless, simplified descriptions of the cure chemistry (in combination with an appropriate cure model) provide reasonable estimates of such important parameters as the progress of the cure and the resin flow. Therefore, simplified descriptions of the chemistry play a useful role in the simulation fo the cure processes of thermosetting resins.

A.1 CHEMICAL KINETICS

The heat of reaction, the rate of cure, and the degree of cure can be measured with a differential scanning calorimeter (DSC). The isothermal heat of reaction H_T is defined as the total amount of heat generated from time $t = 0$ until no evidence is found of further reactions at a constant temperature. The ultimate heat of reaction H_U is the amount of heat generated during dynamic scanning until the completion of the chemical reactions

$$H_T = \int_0^{t_{fT}} \left(\frac{dQ}{dt} \right)_T dt \tag{A.1}$$

$$H_U = \int_0^{t_{fd}} \left(\frac{dQ}{dt} \right)_d dt \tag{A.2}$$

$(dQ/dt)_T$ and $(dQ/dt)_d$ are the instantaneous rates of heat generated during isothermal and dynamic scanning, respectively. t_{fT} and t_{fd} are the amounts of time required to complete the reactions during isothermal and dynamic scannings. The ultimate heat of reaction H_U is a constant, and is independent of the heating rate. H_T depends on the temperature.

The rate of cure is a parameter proportional to the rate of heat release at a constant temperature T

$$\frac{d\alpha}{dt} \equiv \frac{1}{H_U}\left(\frac{dQ}{dt}\right)_T \qquad \text{(A.3)}$$

The degree of cure is proportional to the amount of heat released (at a constant temperature) from time $t = 0$ to time t

$$\alpha \equiv \frac{1}{H_U}\int_0^t \left(\frac{dQ}{dt}\right)_T dt \qquad \text{(A.4)}$$

Note that according to the above definition of α the resin is considered to be uncured when $\alpha = 0$ and completely cured when $\alpha = 1$.

The ultimate heat of reaction H_U is determined directly from the results of dynamic scanning measurements (see Eq. A.2). The values of H_T may be determined either from the results of the isothermal scanning measurements (see Eq. A.1) or from the values of the ratio H_T/H_U. By combining Eqs. A.1 and A.3, we obtain

$$H_T / H_U = [\alpha]_{t \to t fT} \qquad \text{(A.5)}$$

In other words, the ratio H_T/H_U has the value of α when no more heat is released and dQ/dt as well as $d\alpha/dt$ become zero.

The isothermal release rates $(dQ/dt)_T$, the value of H_U, the rate of cure $d\alpha/dt$ and the degree of cure α can be deduced directly from the data generated by DSC. Values of these parameters could be tabulated. Although such tabular representations would suffice for numerical calculations, it is more convenient to use analytical expressions for representing α and $d\alpha/dt$, which are two of the parameters of most interest in cure process simulations.

In order to represent α and $d\alpha/dt$ by analytical expressions, Eq. A.3 is rewritten in the form

$$\frac{d\alpha}{dt} = \frac{H_T}{H_u} \frac{1}{H_T} \left(\frac{dQ}{dt} \right)_T \tag{A.6}$$

or

$$\frac{d\alpha}{dt} = \frac{H_T}{H_u} \frac{d\beta}{dt} \tag{A.7}$$

where $d\beta/dt$ is an "isothermal" rate of cure defined as

$$\frac{d\beta}{dt} \equiv \frac{1}{H_T} \left(\frac{dQ}{dt} \right)_T \tag{A.8}$$

The degree of cure α is obtained by integrating Eq. A.7

$$\alpha = \int_0^t \left(\frac{H_T}{H_U} \right) \frac{d\beta}{dt} dt \tag{A.9}$$

The ratio H_T/H_U is obtained by a least squares curve fit to the experimentally determined values of H_T/H_U versus temperature. Thus $d\alpha/dt$ and α can readily be calculated at any particular temperature from Eqs. A.7 and A.9 once an expression for $d\beta/dt$ is established. It is convenient to assume the following form for $d\beta/dt$

$$\frac{d\beta}{dt} = \left(K_1 + K_2 \beta^m \right) (1 - \beta)^n \tag{A.10}$$

The exponents m and n are constants, independent of temperature. K_1 and K_2 are parameters which depend on the temperature in the following manner

$$K_1 = A_1 \exp(-\Delta E_1 / RT) \tag{A.11}$$

$$K_2 = A_2 \exp(-\Delta E_2 / RT) \qquad \text{(A.12)}$$

A_1 and A_2 are the pre-exponential factors. ΔE_1 and ΔE_2 are the activation energies, R is the universal gas constant, and T is the absolute temperature in degrees Kelvin.

The isothermal degree of cure β corresponding to $d\beta/dt$ is given by

$$\beta = \int_0^t \frac{d\beta}{dt} dt \qquad \text{(A.13)}$$

The constants K_1, K_2, m, and n are determined as follows. First, the values $d\beta/dt$ are calculated by Eq. A.7 using the experimentally determined values of $d\alpha/dt$ and H_T/H_U. Second, the corresponding values of β are calculated using Eq. A.13. Third, the $d\beta/dt$ versus β values are fitted with a least square curve using the Levenberg-Marquardt algorithm. The values of A_1, A_2, ΔE_1, and ΔE_2 are found by fitting straight lines to the K_1, K_2 versus $1/T$ points (see Eqs. A.11 and A.12).

A.2 VISCOSITY

The viscosity can be measured with a viscometer which provides the complex viscosity in poise as a function of time. The complex viscosity is taken to be the same as the shear viscosity. This approximation is reasonable at least at low degrees of cure. It becomes more inappropriate as the gel point is approached. The measured values of the viscosity as functions of temperature and time can be tabulated. The summary of the data in tabular form would be adequate for the purpose of simulating the cure process. However, numerical calculations are greatly facilitated if the viscosity is expressed in analytical form. This can be accomplished by observing that both the viscosity μ and the degree of cure α are functions of temperature and time. Hence a relationship between μ and α can be established. It is convenient to adopt the following form to describe the viscosity

$$\mu = \mu_\infty \exp\left(\frac{U}{RT} + \kappa\alpha\right) \qquad \text{(A.14)}$$

where μ and κ are constants and U is the activation energy for viscosity. As before, R is the universal gas constant and T is the absolute temperature.

The constants μ and κ can be determined by the following procedure. Eq. A.14 can be written as

$$\ln \mu = A + \kappa \alpha \tag{A.15}$$

where A is a parameter defined as

$$A \equiv \ln \mu_{\infty} + U / RT \tag{A.16}$$

The constant κ and the parameter A can be found as follows. First the measured values of μ are plotted versus the calculated values of α (see Eqs. A.7, A.9, and A.10). Note that in this plot, at each point the values of μ and α correspond to the same time and temperature, Second, linear least square curves are fitted to the μ versus α data at each constant temperature. This procedure yields the values of κ in Eq. A.15. The values of exp (A) versus $1/T$ can also be fitted by linear least square curves to find the values of μ and U, yielding the last of the required parameters for Eq. A.14.

APPENDIX B

Calculation of the Prepreg Properties

Solution of the thermo-chemical model described in Chapter 3 requires that the density ρ, specific heat C, heat of reaction H_u, and the thermal conductivity k of the prepreg be known. These properties depend on the local resin and fiber contents of each prepreg ply. The resin content varies in the composite with position and time. Hence ρ, C, H_u, and k also vary with position and time. These properties can be calculated using the rule of mixtures described below.

The mass of the nth prepreg ply is

$$M_n = \left(M_r\right)_n + M_f \tag{B.1}$$

where $(M_r)_n$ and M_f are the masses of the resin and the fiber in the nth prepreg ply, respectively. The mass of the fiber is the same for each prepreg ply, and remains constant during the resin flow process. By expressing the mass in terms of volume and density

$$M = \rho V \tag{B.2}$$

Eq. B.1 can be rewritten as

$$\rho_n V_n = \rho_r (V_r)_n + \rho_f V_f \tag{B.3}$$

where V_n, $(V_r)_n$, and V_f are the volumes of the prepreg, the resin and the fiber, respectively. The volume fraction of the resin and the volume fraction of the fiber can be defined for the nth prepreg ply as

$$(v_r)_n = \frac{(V_r)_n}{V_n} = \frac{(V_r)_n}{(V_r)_n + V_f} \tag{B.4}$$

$$(v_f)_n = \frac{V_f}{V_n} = \frac{V_f}{(V_r)_n + V_f} \tag{B.5}$$

Substitution of Eq. B.2 into Eqs. B.4 and B.5 gives

$$(v_r)_n = \frac{1}{1 + \frac{\rho_r}{\rho_f} \frac{M_f}{(M_r)_n}} \tag{B.6}$$

$$(v_f)_n = \frac{1}{1 + \frac{\rho_f}{\rho_r} \frac{(M_r)_n}{M_f}} \tag{B.7}$$

Equations B.3, B.6, and B.7 yield the following expression for the density

$$\rho_n = \frac{\rho_r}{1 + \frac{\rho_r}{\rho_f} \frac{M_f}{(M_r)_n}} + \frac{\rho_f}{1 + \frac{\rho_f}{\rho_r} \frac{(M_r)_n}{M_f}} \tag{B.8}$$

It should be noted that values of ρ_r and ρ_f can be determined readily. The mass of resin in each prepreg ply $(M_r)_n$ can be calculated from the resin flow models described in Chapter 3.

The specific heat of the nth prepreg ply is taken to be the sum of the specific heat of the resin C_r and the specific heat of the fibers C_f

$$M_n C_n = (M_r)_n C_r + M_f C_f \tag{B.9}$$

By rearranging Eq. B.9, the prepreg specific heat per unit mass C_n can be written as

$$C_n = \frac{(M_r)_n}{M_n} C_r + \frac{M_f}{M_n} C_f \tag{B.10}$$

The heat of reaction for the nth prepreg ply can be expressed as

$$(H_u)_n = \frac{(M_r)_n}{M_n} H_r \tag{B.11}$$

where H_u is the heat of reaction per unit mass for the prepreg and H_r is the heat of reaction per unit mass for the resin. Equation B.11 assumes that the fibers do not participate in the reaction and do not affect the chemical reactions occurring in the resin. In Eqs. B.10 and B.11 the resin and fiber properties are specified and the fiber mass is known for each ply.

The thermal conductivity normal to the fibers (z direction Figure 3.2) may be estimated from the expression [23]

$$(k)_n = \left(1 - 2\sqrt{\frac{(v_f)_n}{\pi}}\right) k_r + \frac{k_r}{B_k}\left[\pi - \frac{4}{\sqrt{1 - \left(\frac{B_k^2 (v_f)_n}{\pi}\right)}} \tan^{-1} \frac{\sqrt{1 - \left(\frac{B_k^2 (v_f)_n}{\pi}\right)}}{1 + B_k\sqrt{\frac{(v_f)_n}{\pi}}}\right] \tag{B.12}$$

where B_k is defined as

$$B_k \equiv 2\left(\frac{k_r}{k_f} - 1\right) \tag{B.13}$$

Here k_r and k_f are the thermal conductivities of the resin and the fibers, respectively. Values of k_r and k_f are known and the fiber volume fraction of the nth prepreg ply can be calculated by Eq. B.7.

SIMULATOR Results

In this appendix, results are presented of cures simulated by the SIMULATOR for the following laminates from Table 5.1.

Material	No. Plys
AS4/3501-6	52
T300/976 Tape	52
(T300/976 Fabric)	(26)

Note that the CURE code does not distinguish between fiber weave styles (i.e. there is no distinction between tape and fabric). The 52 ply T300/976 tape and the 26 ply T300/976 fabric laminate are nearly identical in thickness and hence have nearly identical SIMULATOR results. Therefore, in Figures C.3 and C.4 only the results for a 26 ply T300/976 fabric laminate are given.

Results for the simulated cure of the 16 ply and the 200 ply T300/976 tape laminates are included in the main text (Figures 5.2 - 5.5)

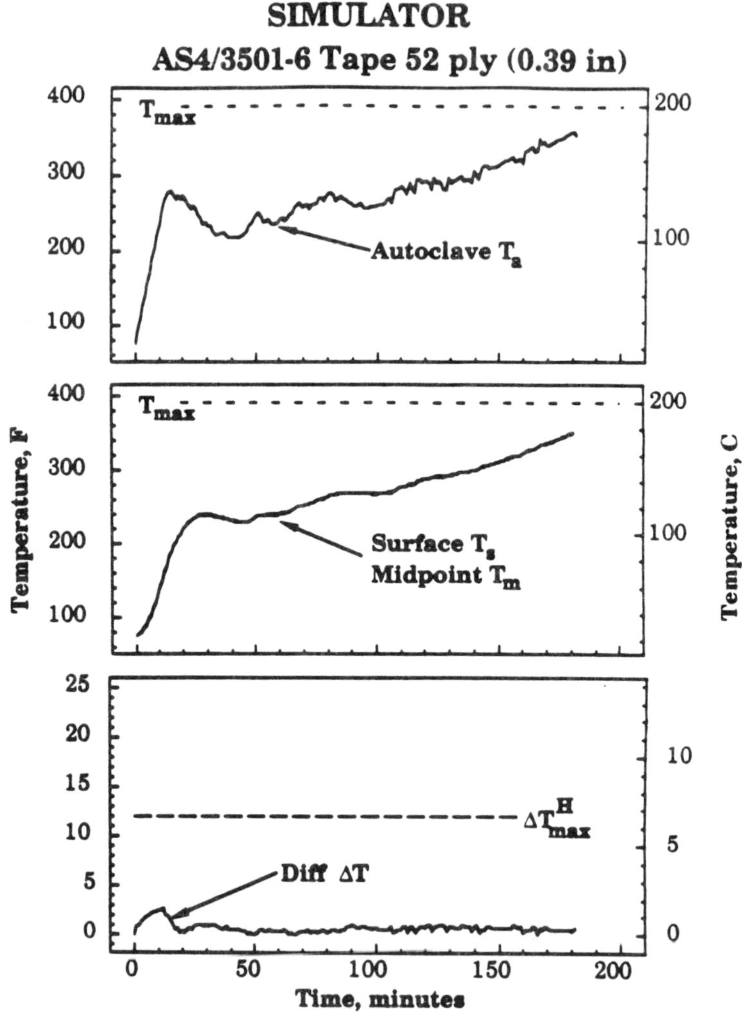

Figure C.1 Temperatures from the SIMULATOR (Table 5.1).

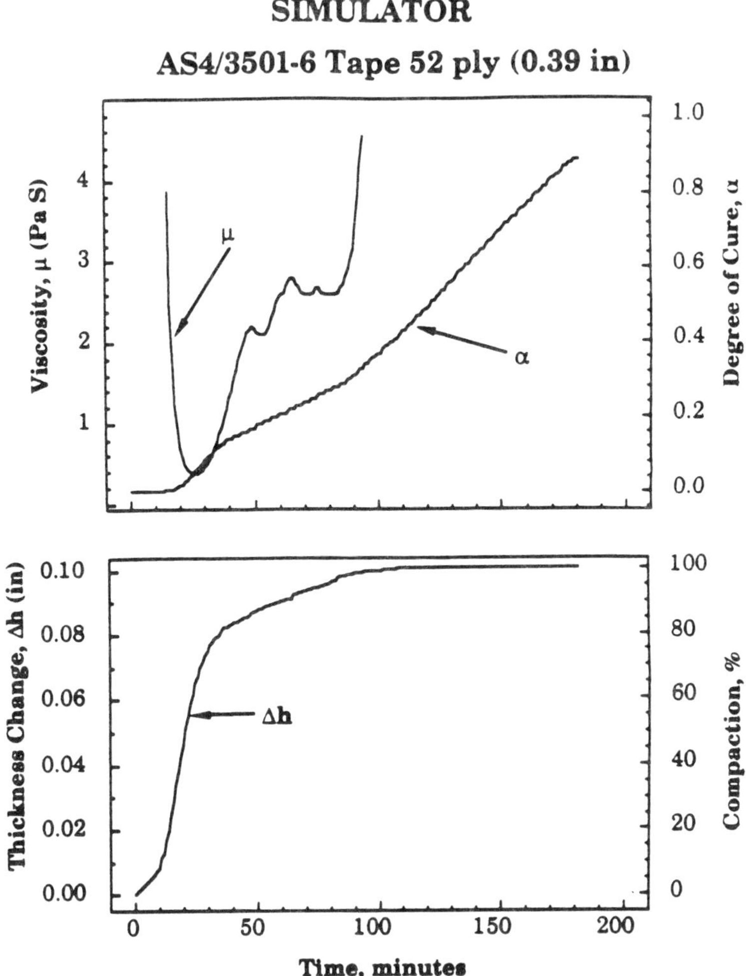

Figure C.2 Viscosity, Degree of Cure and Compaction from the SIMULATOR (Table 5.1).

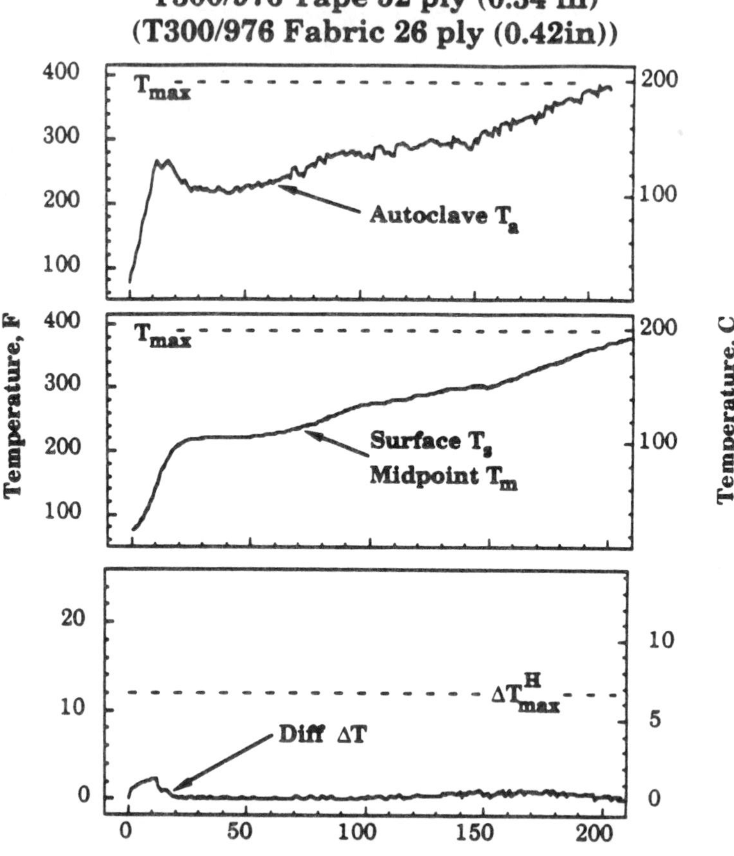

Figure C.3 Temperatures from the SIMULATOR (Table 5.1).

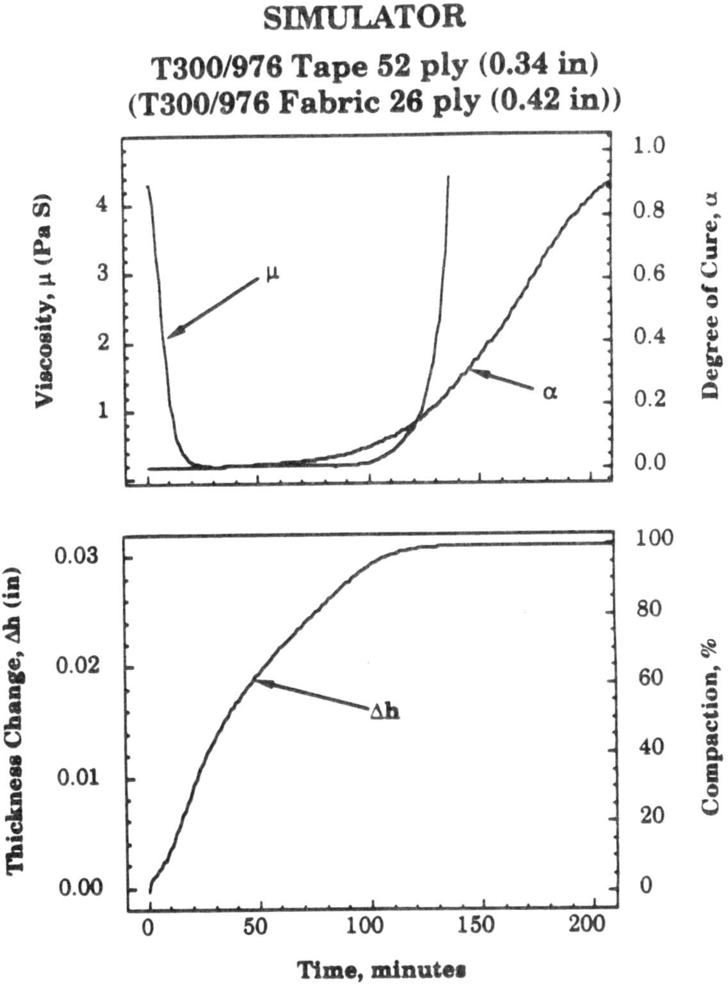
SIMULATOR
T300/976 Tape 52 ply (0.34 in)
(T300/976 Fabric 26 ply (0.42 in))
Viscosity, μ (Pa S)
4
3
2
1
μ
α
1.0
0.8
0.6
0.4
0.2
0.0
Degree of Cure, α
Thickness Change, Δh (in)
0.03
0.02
0.01
0.00
Δh
100
80
60
40
20
0
Compaction, %
0
50
100
150
200
Time, minutes

Figure C.4 Viscosity, Degree of Cure and Compaction from the SIMULATOR (Table 5.1).

Thickness Gauge

The thickness gauge consists of a 0.375 inch diameter, 5 inch long steel rod. The rod is placed inside a 0.430 inch inside diameter, (0.500 in o.d.), 3.5 inch long steel tube which acts as a guide. The tube is attached to a steel frame. Attached to this frame is a (3.5 inch long, 1 inch wide, 0.020 inch thick) spring steel cantilever beam. Two strain gauges (type CEA-060125UN-350, Measurements Group, Incorporated, Raleigh, North Carolina) are mounted on the top and bottom surfaces of this beam near the clamped end. The frame is mounted above the laminate being cured. The tube with the rod inside it is placed between the tip of the cantilever beam and the composite. There is a small (1 x 1 x 0.040 inch) aluminum plate placed on top of the porous film. A hole the size of the plate is cut in the surrounding layers (bleeder, breather, etc.) but not in the vacuum bag. The rod is positioned on top of the vacuum bag at the center of the small aluminum plate. The thickness gauge and the composite layup details are illustrated in Figure D.1.

Vertical movement of the rod results in bending of the cantilever beam and a corresponding change in the strains recorded by the strain gauges. The millivolt output from the gauges are calibrated with respect to displacement of the rod. The strain gauge output is fed to a signal conditioner (described in Appendix E).

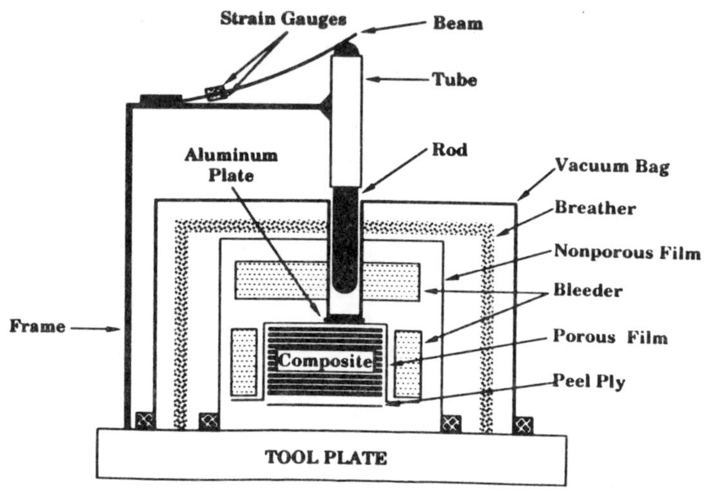

Figure D.1 Schematic of the Thickness Measuring Equipment and Layup of the Composite.

APPENDIX E

Sensor and Control Circuits

A block diagram of the sensor and control circuit is shown in Figure E.1 The four main sensing units which are incorporated into the present cure system are: thermocouples, dielectric, and thickness. Pressure is not included because it is convenient and appropriate to apply the pressure at the start of the manufacturing process. There appears to be no need to change the pressure repeatedly during the cure. One or two necessary pressure changes can best be effected manually. If needed, pressure controllers could easily be incorporated in SECURE in the same manner as the other control units.

The components currently installed in SECURE are described below.

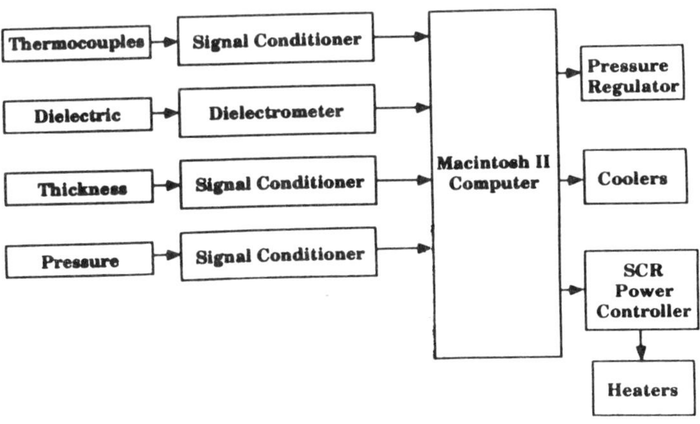

Figure E.1 Block diagram of the sensor and control circuits.

E.1 SENSORS

The millivolt outputs of the thermocouples are fed into a signal conditioner. The circuit diagram of this conditioner is shown in Figure E.2. The output of this conditioner is a 0 - 10 V electric signal which is fed into a Macintosh II computer.

The outputs of the dielectric sensors are connected to a Micromet Incorporated Eumetrics II analyzer. The sensor signals are converted by the instrument to ASCII values representing the permittivity E' and the loss factor E'' and are fed into the serial port of the Macintosh II computer.

The millivolt output of the thickness gauge is fed into a signal conditioner. The circuit diagram of this conditioner is shown in Figure C.3. The output of this conditioner is a 0 - 10 V electric signal which is fed into a Macintosh II computer.

The millivolt outputs of the embedded strain gauges are fed into a signal conditioner. The circuit diagram of this conditioner is the same as that for the thickness gauge (Figure E.3). The output of the strain gauge conditioner is a 0 - 10 V electric signal which is fed into a Macintosh II computer.

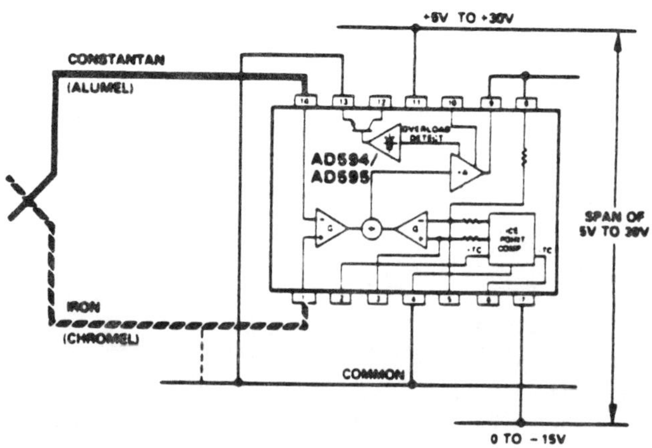

Figure E.2 Diagram of the thermocouple signal conditioner circuit (AD594, Analog Devices Incorporated).

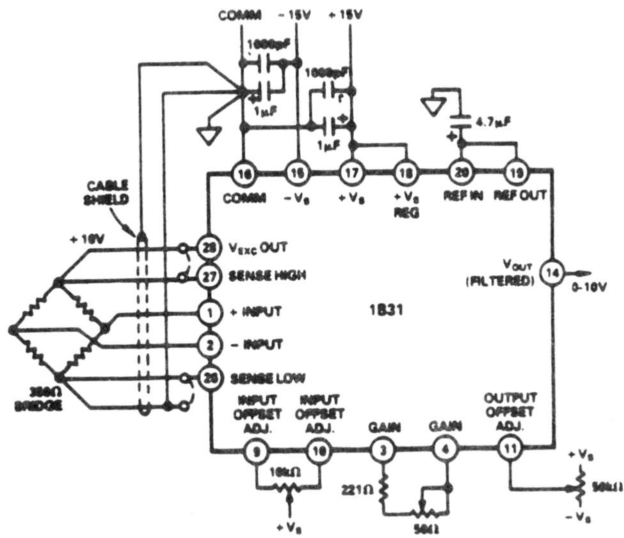

Figure E.3 Diagram of the thickness signal conditioner circuit (1B31, Analog Devices Incorporated)

E.2 COMPUTER

The functions of the computer can be divided into three parts. The "Observer", the "Thinker" and the "Stoker" (Figure E.4). In the Observer the incoming signals are converted to temperature, modified ionic conductivity, change in thickness, and pressure. The voltages corresponding to temperature are converted to temperature using the relationships given by Omega Instruments Corporation and Analog Devices Corporation. The dielectric property values (permittivity E' and loss factor E'') are converted to modified ionic conductivity γ by Eqs. 4.4 and 4.8. The voltage outputs of the thickness gauge, strain gauge and pressure sensor are converted to Δh, and P values obtained by calibration.

In the Thinker, the temperature, ionic conductivity, thickness change, and pressure data are analyzed according to the numerical procedures outlined in section 4.8. The rules are then applied and the indicated decisions are made.

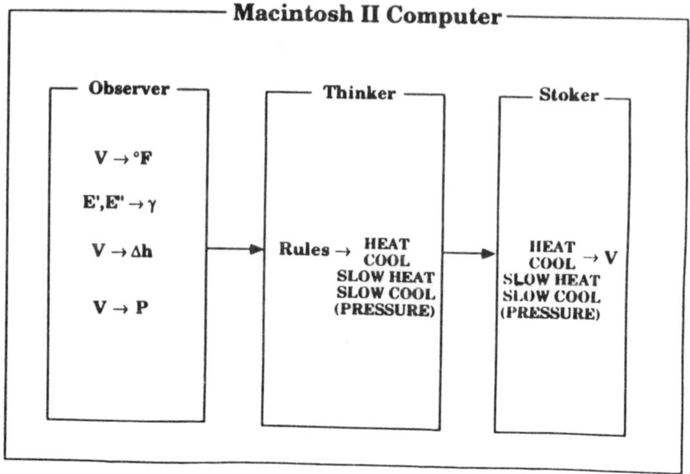

Figure E.4 Block diagram of the information flow in the computer.

The outputs from the thinker are:

a) HEAT
b) SLOW HEAT
c) SLOW COOL
d) COOL

The corresponding character strings are fed into the "Stoker" where they are converted to a voltage on a 0 - 5 volt scale. Zero represents power off to the heaters and maximum power to the coolers. At all other power settings the coolers are off and the heater is on. Five volts represents full power to the heaters.

The intermediate settings of the heaters are determined by calibration. This calibration was done by heating the autoclave to different constant temperatures and by determining at each of these temperatures the amount of heat (power setting) required to raise the autoclave temperature by 1°F.

E.3 AUTOCLAVE CONTROLS

The 0 - 5 volts output from the Macintosh II controller are fed to the autoclave SCR (silicon control rectifier) power controller and to the cooler control. These controls then provide the appropriate power to the autoclave heating elements and cooler controller.

APPENDIX F

Optimized Conventional Cure Data

The temperatures, modified ionic conductivities and compactions measured in laminates cured by optimized conventional cure cycles are presented in this appendix (Figures F.1 - F.5). The values of these parameters calculated by the CURE computer code are also included in these figures (solid lines). In addition, the degree of cure and viscosity values calculated by the CURE code are shown.

From Figures F.1 - F.5, it is seen that the CURE code predicts the temperatures and the compactions with good accuracy. It is further seen that the trend in the modified ionic conductivity is similar to the trend in viscosity.

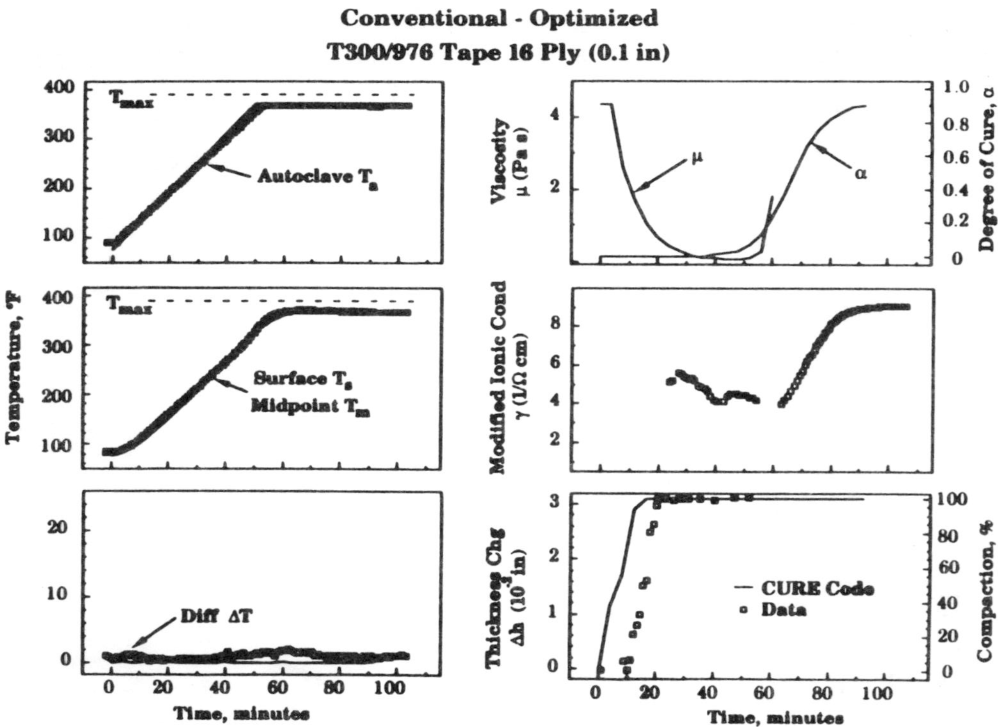

Figure F.1 The temperature, thickness change, compaction, modified ionic conductivity, degree of cure and viscosity for laminates cured by optimized conventional cure (▫ Data - CURE code).

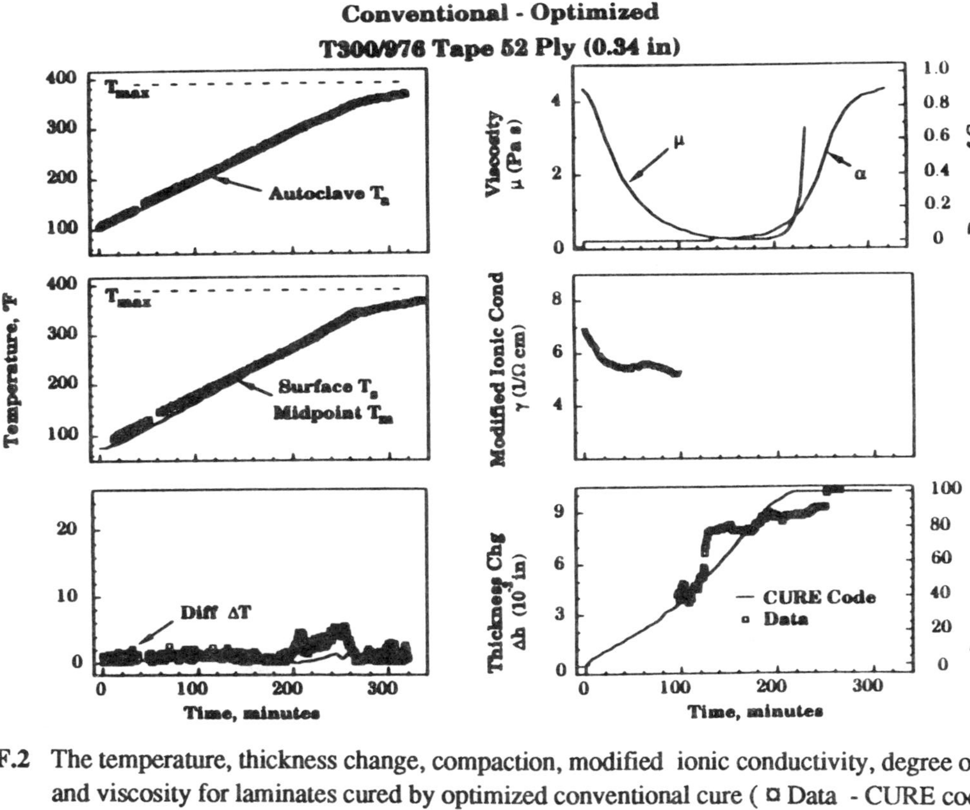

Figure F.2 The temperature, thickness change, compaction, modified ionic conductivity, degree of cure and viscosity for laminates cured by optimized conventional cure (▫ Data - CURE code).

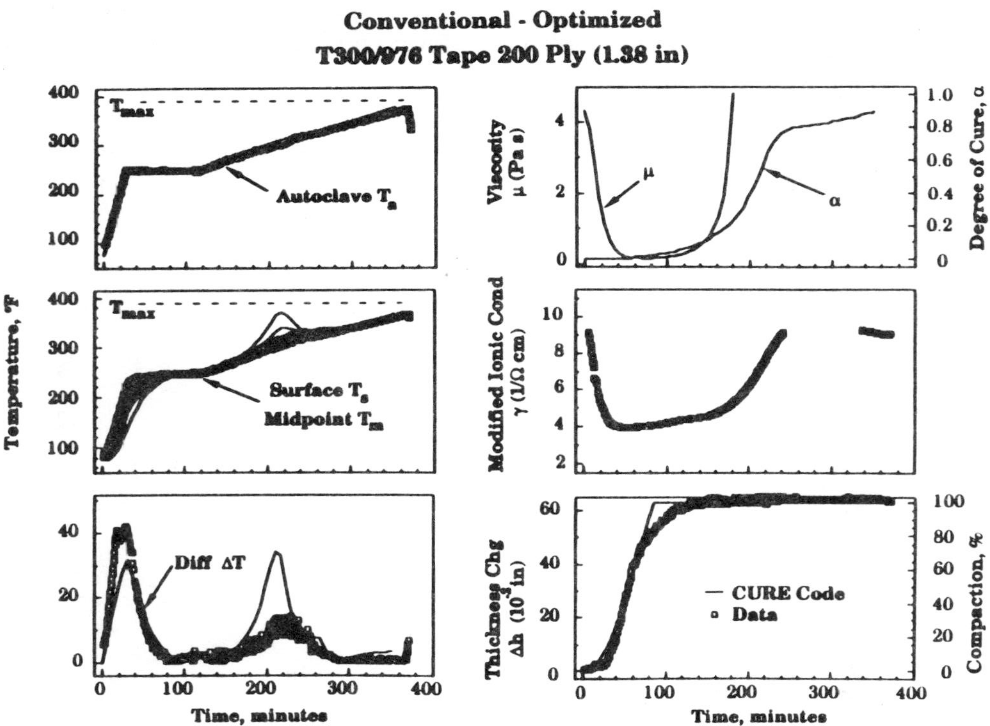

Figure F.3 The temperature, thickness change, compaction, modified ionic conductivity, degree of cure and viscosity for laminates cured by optimized conventional cure (◻ Data - CURE code).

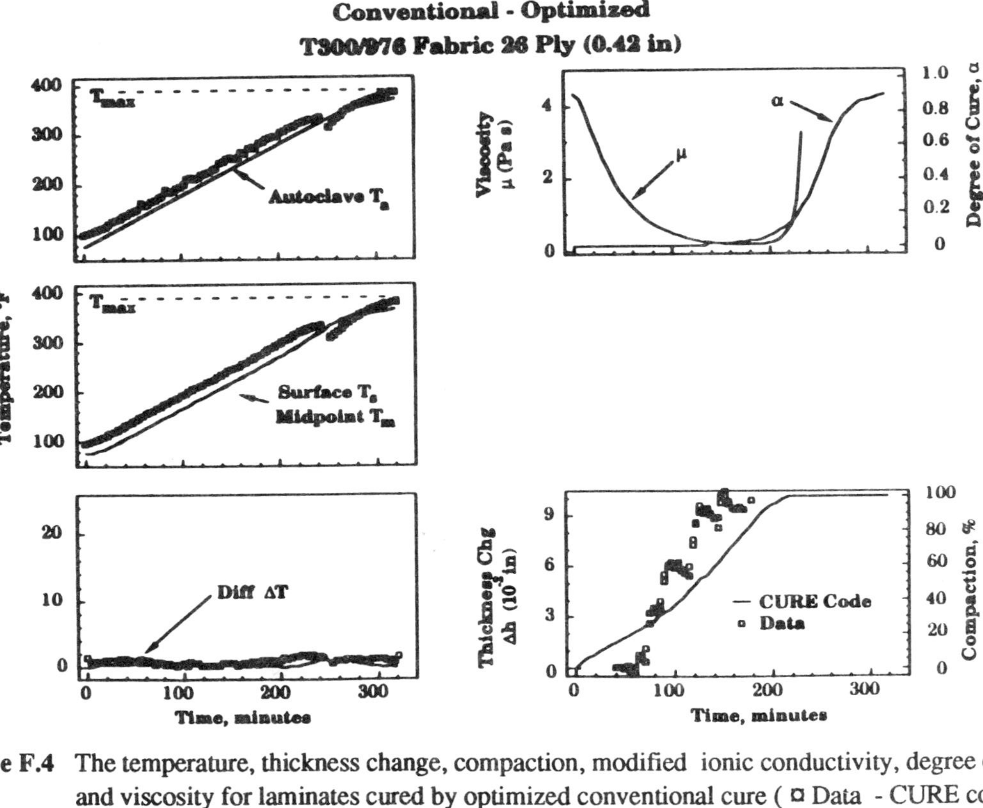

Figure F.4 The temperature, thickness change, compaction, modified ionic conductivity, degree of cure and viscosity for laminates cured by optimized conventional cure (□ Data - CURE code).

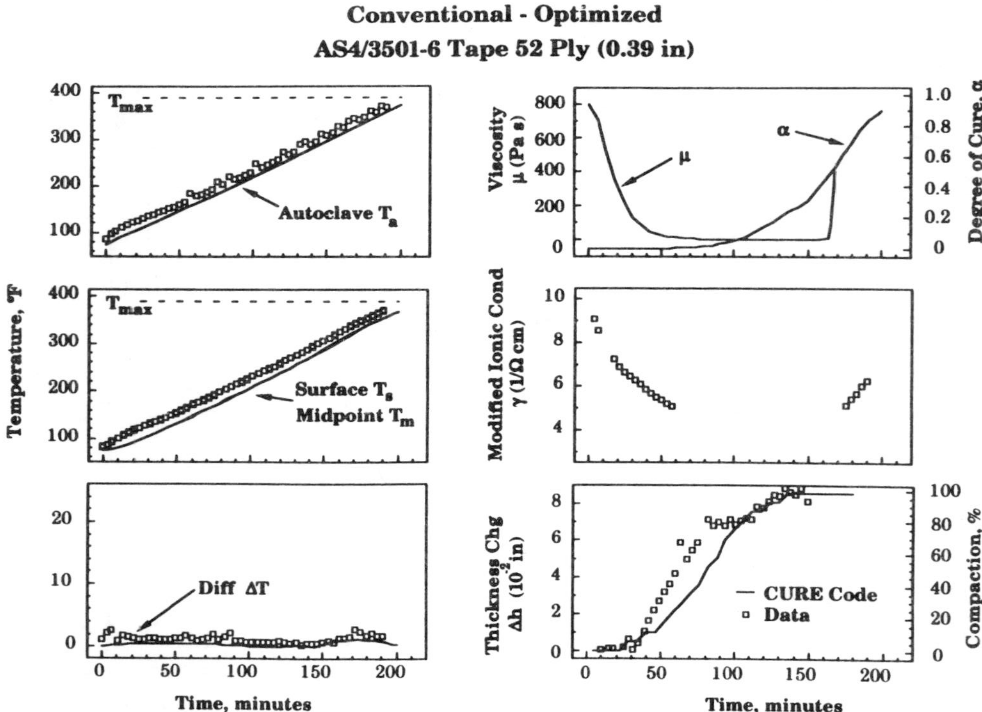

Figure F.5 The temperature, thickness change, compaction, modified ionic conductivity, degree of cure and viscosity for laminates cured by optimized conventional cure (¤ Data - CURE code).

Crystallinity for PEEK 150P Polymer

Several empirical expressoins have been proposed in the past for correlating the measured rate of crystallization with temperature[12, 14-16, 35-39]. The expression proposed by Ozawa [39] is adequate for engineering analysis. Ozawa's expression can be written as

$$\log[-\ln(1-c_r] = \log\phi - n\log\left(\frac{dT}{dt}\right) \tag{G.1}$$

or in differential form

$$\frac{dc_r}{dt} = -(1-c_r)\frac{(d\phi / dT)}{(dT / dt)^{n-1}} \tag{G.2}$$

c_r is the relative crystallinity and is related to the crystallinity by

$$c_r = \frac{H_u}{H_T}c \tag{G.3}$$

c_r can be determined from measurements performed in a differential scanning calorimeter. H_u is the theoretical ultimate heat of crystallization of the polymer. H_T is the total heat of crystallization at the given cooling rate (Figure G.1). ϕ is a parameter which depends on the temperature only and n is a constant. The parameters H_T/H_u, ϕ, and n can be obtained by fitting the model to data obtained by DSC measurements. Some of the relevant properties for PEEK 150P matrix and APC-2 composite are listed in Table G.1.

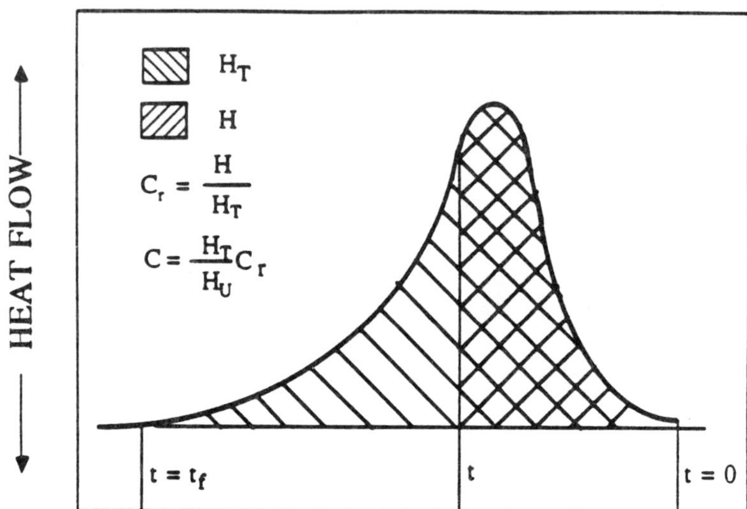

Figure G.1 Illustration of the results obtained on cooling a PEEK 150P polymer sample in the DSC.

Table G.1

Some Properties of the PEEK 150P Matrix and the APC-2 Composite	
PEEK 150P	
Density:	ρ = 81.1 lbm/ft^3
Specific heat:	C = 0.32 BTU/lbm/°F
Thermal conductivity:	k = 0.145 BTU/hr/ft/°F
Ultimate heat of crystallization:	H_u = 0.145 BTU/lbm
APC-2	
Matrix mass fraction:	m = 0.31
Ply thickness:	s_o = 0.005 in

References

1. Springer, G. S. "A Model of the Curing Process of Epoxy Matrix Composites," *Progress in Science and Engineering of Composites*. Eds. T. Hayashi, K. Kawata, and S. Umekawa, *Japan Society for Composite Materials, Vol. 1* (1982), pp. 23 - 35.

2. Loos, A. C. and G. S. Springer. "Curing of Epoxy Matrix Composites," *Journal of Composite Materials, Vol. 17* (1983), pp. 135 - 169.

3. Gutowski, T. G. "A Resin Flow/Fiber Deformation Model for Composites," *SAMPE Quarterly, Vol. 16* (1985), pp. 58 - 64.

4. Gutowski, T. G., T. Morigaki, and Z. Cai. "The Consolidation of Laminate Composites," *Journal of Composite Materials, Vol. 21* (1987), pp. 172 - 188.

5. Dave, R., J. L. Kardos, and M. P. Dudukovic. "A Model for Resin Flow During Composite Processing. Part 1: General Mathematical Development," *Polymer Composites, Vol. 8* (1987), pp. 29 - 38.

6. Kardos, J. L., R. Dave, and M. P. Dudukovic. "Void Growth and Resin Transport During Processing of Thermosetting Matrix Composites," Advances in Polymer Science, Vol. 80 Part IV. New York: SpringerVerlag, 1986, pp. 101 - 123.

7. Campbell, F. C., A. R. Mallow, F. R. Muncaster, B. L. Boman, and G. L. Blase. "Computer Aided Curing of Composites," Air Force Materials Laboratory Report, AFWAL-TR-86-4060, Dayton Ohio, 1986.

8. Lee, S. Y. and G. S. Springer. "Filament Winding Cylinders, I: Process Model," *Journal of Composite Materials*, (1990).

9. Calius E. P., S. Y. Lee and G. S. Springer. " Filament Winding Cylinders, II: Validation of the Process Model," *Journal of Composite Materials*, (1990).

10. Lee, S. Y. and G. S. Springer. " Filament Winding Cylinders, III: Selection of the Process Variables," *Journal of Composite Materials*, (1990).

11. Calius E. P. and G. S. Springer. " A Model of Filament Wound Thin Cylinders," *International Journal of Solids and Structures, Vol. 26* (1989), pp. 271 - 297.

12. Lee, W. I. and G. S. Springer. " A Model of the Manufacturing Process of Thermoplastic Matrix Composites," *Journal of Composite Materials, Vol. 21* (1987), pp. 1017 - 1055.

13. Dara, P. H. and A. C. Loos. " Thermoplastic Matrix Composite Processing Model," Center for Composite Materials and Structures, Report CCMS-85-10, VPI-E-85-21, Virginia Polytechnic Institute and State University, Blacksburg, Virginia, 1985.

14. Lee, W. I., M. F. Talbott, G. S. Springer and L.A. Berglund. "Effect of Cooling Rate on the Crystallinity and Mechanical Properties of Thermoplastic Composites," *Journal of Reinforced Plastics and Composites, Vol. 6* (1987), pp. 2 - 12.

15. Seferis, J. C. and C. N. Velisaris. "Modeling-Processing-Structure Relationships of Polyetheretherketone (PEEK) Based Composites," in *Materials for the Future*, Society for the Advancement of Material and Process Engineering, Vol. 31, (1986), pp. 1236 - 1252.

16. Velisaris, C. N. and J. C. Seferis. "Crystallization Kinetics of Polyetheretherketone (PEEK) Matrices," *Polymer Engineering and Science, Vol. 26* (1986), pp. 1574 - 1581.

17. Abrams, F., P. Garrett, T. Lagnese, S. LeClair, C. W. Lee and J. Park. "Qualitative Process Automation of Autoclave Curing of Composites," Air Force Materials Laboratory Report, AFWAL-TR-87-4083, Dayton, Ohio, 1987.

18. Servais, R. A., C. W. Lee and C. E. Browning. "Intelligent Processing of Composite Materials," in *Materials for the Future*, Society for the Advancement of Material and Process Engineering, Vol. 31, (1986), pp. 764 - 775.

19. Ciriscioli, P. R. "An Expert System for Autoclave Curing Composites," Ph.D. Thesis, Stanford University, 1990.

20. Lee, W. I., A. Loos and G. S. Springer. "Heat of Reaction, Degree of Cure and Viscosity of Hercules 3501-6 Resin," *Journal of Composite Materials, Vol. 16* (1982), pp. 510 - 520.

21. Dusi, M. R., W. I. Lee, P. R. Ciriscioli and G. S. Springer. "Cure Kinetics and Viscosity of Fiberite 976 Resin," *Journal of Composite Materials, Vol. 21* (1987), pp. 243 - 261.

22. Springer, G. S. "Curing of Graphite/Epoxy Composites," Air Force Materials Laboratory Report AFWAL-TR-4040, Dayton Ohio, 1983.

23. Springer, G. S. and S. W. Tsai. "Thermal Conductivities of Unidirectional Materials," *Journal of Composite Materials, Vol. 1* (1967), pp. 166 - 173.

24. Springer, G. S. "Resin Flow During the Cure of Fiber Reinforced Composites," *Journal of Composite Materials, Vol. 16* (1982), pp. 400 - 410.

25. Bartlett, C. J. "Use of the Parallel Plate Plastometer to Characterize Glass-Reinforced Resins: I. Flow Model," *SPE Technical Papers, Vol. 24* (1978), pp. 638 - 640.

26. White, F. M. "Viscous Fluid Flow," McGraw-Hill, 1974.

27. Kardos, J. L., J. P. Dudukovic, E. L. McKague, and M. W. Lehman. "Void Formation and Transport During Composite Laminate Processing," in *Composite Materials, Quality Assurance and Processing*, (C. E. Browning, ed.), ASTM STP 797, 1983, pp. 96 - 109.

28. Springer, G. S. "Environmental Effects on Composite Materials," Technomic Publishing Co., 1981.

29. Tsai, S. W. and H. T. Hahn. "Introduction to Composite Materials," Technomic Publishing Co., 1980.

30. Jones, R. M. "Mechanics of Composite Materials," McGraw-Hill, 1980.

31. "Advanced Composite Processing Technology Development," Contract F33615-88-C-5455, McDonnell Douglas Corporation Third Quarterly Interim Report, St. Louis, Missouri, 1989.

32. Ciriscioli, P. R. and G. S. Springer. "Dielectric Cure Monitoring - A Critical Review," *SAMPE Journal, Vol. 25* (1989), pp. 35 - 42.

33. Kim, T. W., E. J. Jun and W. I. Lee. "The Effect of Pressure on the Impregnation of Fibers with Thermoplastic Resin," in *Tomorrow's Materials Today*, Society for the Advancement of Material and Process Engineering, Vol. 34, (1989), pp. 323 - 328.

34. Eckert, E. R. and R. M. Drake. "Heat and Mass Transfer," McGraw-Hill, 1975.

35. Blundell, D. J., and B. N. Osborn. "The Morphology of Poly(aryl-ether-etherketone)," *Polymer, Vol. 24* (1983), pp. 953 - 958.

36. Blundell, D. J., J. M. Chalmers, M. W. MacKenzie and W. F. Gaskin. "Crystalline Morphology of the matrix of PEEK-Carbon Fiber Aromatic Polymer Composites. I. Assesment of Crastallinity," *SAMPE Quarterly, Vol. 16* (1985), pp. 22 - 30.

37. Blundell, D. J., and B. N. Osborn. "Crystalline Morphology of the matrix of PEEK-Carbon Fiber Aromatic Polymer Composites. II. Crystallization Behavior," *SAMPE Quarterly, Vol. 17* (1985), pp. 1 - 6.

38. Cebe, P. and S. D. Hong. "Crystallization Behavior of Poly(ether-ether-ketone)," *Polymer, Vol. 27* (1986), pp. 1183 - 1192.

39. Ozawa, T. "Kinetics of Nonisothermal Crystallization," *Polymer, Vol. 12* (1971), pp. 150 - 158.

Index